**Power
Generation and
Environmental
Change**

The MIT Press,
Cambridge, Massachusetts,
and London, England

Symposium of the
Committee on
Environmental Alteration,
American Association for
the Advancement of Science,
December 28, 1969

Power Generation and Environmental Change

Edited by
David A. Berkowitz
and Arthur M. Squires

Copyright © 1971 by
The Massachusetts Institute of
Technology

Set in Linotype Baskerville by
The Colonial Press Inc.
Printed by Halliday Lithograph Corp.
Bound in the United States of
America by The Colonial Press Inc.

Reproduction of Chapters 9, 18, 19,
and 21 is permitted for any purpose
of the United States Government.

All rights reserved. No part of this
book with the exception noted
may be reproduced in any form
or by any means, electronic or
mechanical, including photocopying,
recording, or by any information
storage and retrieval system, without
permission in writing from the
publisher.

ISBN 0 262 02072 6 (hardcover)

Library of Congress catalog card
number: 70-137468

Contents

Preface   x

Committee on Environmental Alteration   xiv

Symposium Program   xvi

List of Contributors   xviii

I
Power, Man, and Environment

1 Environmental Science and Public Policy   James A. Fay   3

2 Power Generation and Human Ecology   Frederick E. Smith   7

3 Future Needs for Power from Coal   Wallace B. Behnke, Jr.   11

II
Nuclear Power and Radionuclides in the Environment

4 Nuclear Power Reactors and the Radioactive Environment   Merril Eisenbud   23

Contents

**5
Nuclear Reactors and the
Public Health and Safety**
Arthur Tamplin
44

**6
Other Views on
Public Health and Safety**

I Introduction
The Editors
61

II Comments on
"Nuclear Reactors and the
Public Health and Safety"
Charles W. Edington,
John R. Laughnan,
George E. Stapleton,
and Daniel W. Wilson
63

III Comments on
"Nuclear Reactors and the
Public Health and Safety"
Merril Eisenbud
75

IV Low-Level Radiation
and Health of the Public
Shields Warren
78

**7
Investigation of the
Effects of X-Ray Exposure
of Human Female Fetuses
as Measured by Later
Reproductive Performance:
Interim Summary**
Mary B. Meyer,
Earl L. Diamond,
and Timothy Merz
82

**8
Radiation Dose Limits**
Study Panel on
Nuclear Power Plants,
Maryland Academy of
Sciences
94

Contents

9
Radiation Doses from Fossil-Fuel and Nuclear Power Plants — James E. Martin, Ernest D. Harward, and Donald T. Oakley — 107

Discussion Questions for Part II — 126

III
Hydroelectric Power

10
Ecological Effects of Hydroelectric Dams — Karl F. Lagler — 133

11
Pumped Storage Hydroelectric Projects — David A. Berkowitz — 158

IV
Fossil-Fuel Power

12
Clean Power from Coal, at a Profit — Arthur M. Squires — 175

13
Dealing with Sulfur in Residual Fuel Oil — Seymour B. Alpert, Ronald H. Wolk, and Arthur M. Squires — 228

14
Climatic Consequences of Increased Carbon Dioxide in the Atmosphere — Gordon J. F. MacDonald — 246

15
Atmospheric Chemistry   Richard D. Cadle   263
                                and Eric R. Allen

16
The Fate of SO_2 and   Erik Eriksson   289
NO_x in the Atmosphere

17
An Isotope-Ratio Method   Meyer Steinberg   302
for Tracing Atmospheric
Sulfur Pollutants

18
Environmental Aspects   Harry Perry   317
of Coal Mining

V
Waste Heat

19
Environmental Quality and   S. Fred Singer   343
the Economics of Cooling

20
Impact of Waste Heat   Clarence A. Carlson, Jr.   351
on Aquatic Ecology

21
Thermal Effects—   Walter G. Belter   365
a Potential Problem
in Perspective

22
**Comments on the Use
and Abuse of Energy
in the American Economy**   Robert T. Jaske   387

23
**Alternative Technologies
for Discharging Waste Heat**   William H. Steigelmann   394

Glossary of Nuclear Terms   412
and Units

Index   415

Preface

This book comprises the proceedings of the symposium on Power Generation and Environmental Change with additional, invited contributions. The symposium was part of the 136th Annual Meeting of the American Association for the Advancement of Science. It was sponsored by the AAAS Committee on Environmental Alteration and arranged by Arthur M. Squires, a member of the committee, and David A. Berkowitz, the executive secretary.

AAAS annual meetings have become complex affairs, with many simultaneous symposia of great interest and timeliness. It is difficult to maintain continuity of subject in a single symposium that lasts more than one day, and unreasonable to expect uniform repeat attendance. The symposium on Power Generation and Environmental Change was limited to a single day to encourage all attendees to stay and hear the whole story. It was long and exhausting but appeared to be successful.

In preparing the proceedings, the editors offered the participants the opportunity to contribute papers that were longer and more complete than those they were able to deliver at the symposium. Some participants abbreviated their remarks to permit more discussion and to help keep to a rather tight time schedule. For those who wish a record of what was said, and who prefer the flavor of actual presentations, tape recordings of the entire symposium and discussion periods are available from the AAAS. This volume represents a more complete exposition of the subject matter and reflects more accurately what the participants wish to have available for future reference.

The arrangement of the book parallels that of the symposium. The main subject areas are nuclear power, hydroelectric power, and fossil-fuel power. Waste heat is a product of both nuclear and fossil-fuel power plants. At the symposium, there was a panel discussion on the subject; here the panelists have contributed papers for a separate part at the end of the book.

The editors invited additional contributions to deal with questions raised in discussion periods, points of disagreement,

Preface   xi

and areas not planned for the symposium because of time restrictions. We wish here to identify each added contribution and to express our appreciation to the authors who responded so rapidly in the preparation of their papers.

1.

James A. Fay was chairman of the morning session and introduced the subject matter of the symposium to the attendees. He contributed an introduction for this book (Chapter 1).

2.

Charles W. Edington, John R. Laughnan, George E. Stapleton and Daniel W. Wilson (jointly), Merril Eisenbud, and Shields Warren contributed additional views and discussion on the subject matter of Chapter 5 — health and safety aspects of nuclear radiation. Their contributions comprise Chapter 6.

3.

The possibility of long-term genetic effects was introduced at the symposium, and recent work at The Johns Hopkins University was cited. The work was unfamiliar to many participants and attendees, and subsequent discussion was necessarily incomplete. The group performing the work was invited to provide a more complete statement, which has become Chapter 7, by Mary B. Meyer, Earl L. Diamond, and Timothy Merz.

4.

Chapter 8 is an independent look at the meaning of the radiation-dose limits. It is taken from a report to the Maryland Academy of Sciences on the siting of a nuclear power plant on Chesapeake Bay.

5.

Fossil-fuel power plants emit radioactivity which derives from that contained in the unburned fuel. The nuclear industry occasionally tweaks the fossil-fuel industry about that. James E. Martin, Ernest D. Harward, and Donald T. Oakley have modified a study report of theirs in order to clarify this issue (Chapter 9).

6.
Large pumped storage hydroelectric plants are relatively new to the power economy, and unfamiliar to many environmentalists. Chapter 11, by David A. Berkowitz, describes this form of hydroelectricity and its environmental consequences.

7.
Although coal remains the major energy resource for power in the United States, fuel oil is important in urban areas and is the primary fuel in many other countries. Seymour B. Alpert, Ronald H. Wolk, and Arthur M. Squires have added Chapter 13 to consider the question of sulfur in oil.

8.
Finally, since so many of the environmental aspects of power generation involve pollution of the atmosphere, the editors felt it appropriate to provide a tutorial on atmospheric chemistry. Chapter 15, by Richard D. Cadle and Eric R. Allen, which originally appeared in *Science,* discusses the mechanisms of reaction that occur in clean and dirty air.

The discussion questions at the end of some chapters are not the same as those on the tape recording of the symposium. The ones that appear here were submitted to the editors by the attendees on a written form. They were selected, edited, and referred to the appropriate participant to prepare an answer that became part of his contribution to this book.

The time frame for most of the papers is the next twenty to thirty years. During the symposium the year 2000 was often mentioned, and participants were willing to consider the problems of that year. Few could contemplate the year 2010, and it became clear that not enough thinking has been done about problems that will arise when mankind finally realizes it must curb its desire for ever-increasing amounts of power and move toward the establishment of an economy in which a fixed quantity of power is allocated to uses having the greatest social value.

The editors recognize that the chapters in this book do not provide satisfying answers to crucial long-range questions: What is the practical (or desirable) upper limit for power generation on this planet? And what social arrangements must be made to distribute fairly a fixed amount of power? We can only state that the work necessary before these questions can be tackled has not yet been done. The AAAS Committee on Environmental Alteration has organized a task force to seek answers to the questions and will report to the AAAS annual meeting in December 1971.

Since the time frame for the book did not extend into the twenty-first century, the editors did not feel a need for separate chapters dealing explicitly with power-generation technologies which may become important in that century but which can hardly be expected to become important in this one. Some reference to such techniques as breeder nuclear reactors, magnetohydrodynamic generators, and fuel cells may be traced through the index. The omission of explicit discussions of these techniques should not be viewed by advocates of their development to reflect the editors' judgment of their importance.

There were two deviations from the symposium program as listed. The paper by Harry Perry, who was unable to attend, was read by William L. Crentz, Director of Coal Research for the U.S. Bureau of Mines. Clarence A. Carlson, Jr., was not able to leave Ithaca; the airport and almost everything else was snowed in. S. Fred Singer summarized his paper during the panel discussion.

The editors thank all contributors for their timely response in preparing manuscripts and their enthusiastic participation. We appreciate the encouragement and suggestions of Jack P. Ruina, Chairman of the AAAS Committee on Environmental Alteration, and the committee members in planning the symposium. The patient guidance of Walter G. Berl and D. W. Thornhill of the AAAS in coordination and arrangements was very helpful. Finally, we wish to acknowledge the services of Madeline Pestana of The MITRE Corporation in preparing the manuscript.

David A. Berkowitz
Arthur M. Squires

**Committee on
Environmental Alteration**

American Association for the
Advancement of Science

Jack P. Ruina
Chairman

Massachusetts Institute
of Technology,
Cambridge, Massachusetts

T. C. Byerly

U.S. Department
of Agriculture,
Washington, D.C.

John E. Cantlon

Michigan State University,
East Lansing, Michigan

William M. Capron

Harvard University,
Cambridge, Massachusetts

Barry Commoner

Washington University,
St. Louis, Missouri

H. Jack Geiger

Tufts University
School of Medicine,
Boston, Massachusetts

Jacob E. Goldman

Xerox Corporation,
Rochester, New York

Oscar Harkavy

Ford Foundation,
New York, New York

Walter Modell

Cornell University
Medical College,
New York, New York

Arthur M. Squires

City College of the
City University of New York,
New York, New York

Committee on Environmental Alteration

Dael Wolfle   ex officio, AAAS

David A. Berkowitz   The MITRE Corporation,
Executive Secretary   Bedford, Massachusetts

William T. Kabisch   staff representative, AAAS

Symposium Program   December 28, 1969
                                Boston, Massachusetts

Power Generation and   Reconciling Man's
Environmental Change   Desire for Power
                                with the Needs of
                                His Environment

Morning Session   James A. Fay, Chairman

Nuclear Reactors and the
Radiation Environment   Merril Eisenbud

Man-Made Radiation
in the Biosphere   Arthur Tamplin

Technological Means for
Controlling Radionuclides   Floyd L. Culler

Environmental Change in
Dam Redevelopment Regions   Karl F. Lagler

Future Needs for Power
from Coal   Wallace B. Behnke, Jr.

Environmental Aspects
of Coal Mining   Harry Perry

Afternoon Session   Jack P. Ruina, Chairman

Atmospheric CO_2 Gordon J. F. MacDonald

Fate of SO_2 and NO_x
in the Atmosphere   Erik Eriksson

Isotope Ratio Method
for Tracing Atmospheric
Sulfur Pollutants   Meyer Steinberg

Technological Means for
Controlling SO_2 Arthur M. Squires

Panel Discussion   S. Fred Singer, Chairman
Alternative Technologies for   Walter Belter
Discharging Waste Heat: Clarence A. Carlson, Jr.
Dollar and Environmental   Robert T. Jaske
Costs   Frederick E. Smith
                                       William H. Steigelmann

List of Contributors

Eric R. Allen
Head of the Photochemistry Program

National Center for Atmospheric Research
(Sponsored by the National Science Foundation)
Boulder, Colorado

Seymour B. Alpert
Manager of Development

Hydrocarbon Research, Inc.
Trenton, New Jersey

Wallace B. Behnke, Jr.
Vice-President

Commonwealth Edison Company
Chicago, Illinois

Walter G. Belter
Chief of Environmental and Sanitary Engineering Branch

Division of Reactor Development and Technology
U.S. Atomic Energy Commission
Washington, D.C.

David A. Berkowitz
Coeditor

The MITRE Corporation
Bedford, Massachusetts
(Former Executive Secretary, AAAS Committee on Environmental Alteration)

Richard D. Cadle
Head of the Department of Chemistry

National Center for Atmospheric Research
(Sponsored by the National Science Foundation)
Boulder, Colorado

List of Contributors

Clarence A. Carlson, Jr.
Assistant Professor of Fishery Biology

Assistant Leader, New York Cooperative Fishery Unit
Cornell University
Ithaca, New York

Earl L. Diamond
Professor

Department of Epidemiology
School of Hygiene and Public Health
The Johns Hopkins University
Baltimore, Maryland

Charles W. Edington
Chief of Biology Branch

Division of Biology and Medicine
U.S. Atomic Energy Commission
Washington, D.C.

Merril Eisenbud
Professor of Environmental Medicine and Director

Laboratory for Environmental Studies
Institute of Environmental Medicine
New York University Medical Center
New York, New York
(Formerly Administrator, Environmental Protection Administration, City of New York)

Erik Eriksson
Professor

University of Stockholm
International Meteorological Institute
Stockholm, Sweden

List of Contributors

James A. Fay
Professor of Mechanical
Engineering

Massachusetts Institute of
Technology
Cambridge, Massachusetts

Ernest D. Harward
Deputy Director

Division of Environmental
Radiation
Bureau of Radiological Health
U.S. Department of Health,
Education, and Welfare
Washington, D.C.

Robert T. Jaske
Research Associate

Water Resources Systems
Battelle Northwest
Richmond, Washington

Karl F. Lagler
Professor of Fisheries and of
Zoology

School of Natural Resources
The University of Michigan
Ann Arbor, Michigan

John R. Laughnan
Professor in the Department
of Botany

University of Illinois
Urbana, Illinois
(Formerly at Division of
Biology and Medicine,
U.S. Atomic Energy
Commission,
Washington, D.C.)

Gordon J. F. MacDonald
Council on Environmental
Quality

Executive Office of the
President
Washington, D.C.
(Formerly Vice-Chancellor for
Research and Graduate
Affairs, University of
California at Santa Barbara)

List of Contributors

James E. Martin
Chief of Nuclear
Facilities Branch

Division of Environmental
Radiation
Bureau of Radiological Health
U.S. Department of Health,
Education, and Welfare
Washington, D.C.

Timothy Merz
Associate Professor

Department of Radiological
Science
School of Hygiene and Public
Health
The Johns Hopkins University
Baltimore, Maryland

Mary B. Meyer
Assistant Professor

Department of Epidemiology
School of Hygiene and Public
Health
The Johns Hopkins University
Baltimore, Maryland

Donald T. Oakley
Division of Environmental
Radiation

Bureau of Radiological Health
U.S. Department of Health,
Education, and Welfare
Washington, D.C.

Harry Perry
Senior Specialist in
Environmental Policy

Legislative Reference Service
Library of Congress
Washington, D.C.
(Formery Research Adviser,
Office of the Assistant Secretary
for Mineral Resources, U.S.
Department of the Interior,
Washington, D.C.)

List of Contributors

S. Fred Singer
Deputy Assistant Secretary of the Interior for Scientific Programs

U.S. Department of the Interior
Washington, D.C.

Frederick E. Smith
Professor of Resources and Ecology

Graduate School of Design
Harvard University
Cambridge, Massachusetts

Arthur M. Squires
Coeditor

Professor and Chairman
Department of Chemical Engineering
The City College of the City University of New York
New York, New York

George E. Stapleton
Division of Biology and Medicine

U.S. Atomic Energy Commission
Washington, D.C.

William H. Steigelmann
Manager, Heat and Fluid Mechanics Laboratory

Franklin Institute Research Laboratories
Philadelphia, Pennsylvania

Meyer Steinberg
Supervisor

Radiation Processing Section
Department of Applied Science
Brookhaven National Laboratory
Upton, New York

Arthur Tamplin
Biomedical Division

Lawrence Radiation Laboratory
University of California
Livermore, California

List of Contributors

Shields Warren, M.D.
Cancer Research Institute

New England Deaconess
Hospital
Boston, Massachusetts

Daniel W. Wilson
Division of Biology and
Medicine

U.S. Atomic Energy
Commission
Washington, D.C.

Ronald H. Wolk
Pilot Plant Manager

Hydrocarbon Research, Inc.
Trenton, New Jersey

I

Power, Man,
and Environment

1

Environmental Science and Public Policy James A. Fay

The exponential growth of cheap electric power has vastly increased man's industrial productivity and the amenities of daily living. But this same growth has also intensified deleterious side effects we can no longer ignore: the release into the atmosphere of poisonous or radioactive gases, the heating and mixing of rivers and lakes, the pollution of natural bodies of water from the mining of coal and oil, and the scarring of the landscape by pathways for transmission lines.

Until recently it was commonly assumed that the air, water, and earth had miraculous powers of regeneration and recuperation. Each community was an open system through which flowed streams of clean air and water, sweeping away pollutants that were subsequently rendered harmless by natural processes. We have come to see the inadequacies of this view when it is applied to areas of intense industrial or agricultural activity. Indeed, there are those who look further ahead in time and longer in scope, who see the earth as a closed system with finite amounts of air, water, and arable land and limited rates of processing chemical and biological species. According to this broader view, the extrapolation into the future of our present rate of growth of resource consumption portends disaster.

In the United States the production of electric power has doubled every ten years for the past three decades. It is confidently predicted that it will double again by 1980, most likely double that by 1990, and probably double again by 2000 or shortly thereafter. What will be the effects on the environment of this eightfold increase in power production, which many of us will live to see? Are there technological fixes that will ameli-

The references for Chapter 1 are on page 6.

orate the environmental damage? What will limit this prodigious growth: the supply of fuel, the satiation of the consumer, or the loss of an amenity environment? Indeed, what would a steady-state world be like? Some of these questions are discussed in this volume.

There are two general approaches to solving the environmental problem. The first proposes that technological improvements reduce environmental hazards to a tolerable level and that whatever permanent environmental alteration ensues be a cost that society is willing to pay for the material gains that it will receive from increased utilization of energy and materials. The second approach suggests that per capita consumption, and even population itself, should be controlled or rationed to achieve a steady-state, ecologically balanced civilization in which technological advance might affect the quality of life but not the quantity of material goods. Although public opinion today favors the first view, a small but vociferous minority supports the latter. The distinction is not now a matter for scientific debate but is likely to become so as more and more irreversible changes in the environment are observed or predicted. Most of the chapters in this book, however, are concerned with the means whereby the first approach can be attained, at least for the near future.

The deterioration of the environment has become a national political issue. Many scientists experienced in environmental problems are shocked when angry citizens, demanding an environmental cleanup, denigrate the scholars' expertise and impugn their motives. These scientists sometimes suggest that such confrontations are the result of an ill-informed public which is manipulated by crackpot conservationists or, worse yet, traitorous fellow scientists of questionable reputation. The Nobel laureate Glenn Seaborg, who is also the chairman of the Atomic Energy Commission, has stated that concern for environmental effects "has engendered much irrational thinking and activity based on misinformation and unfounded fears. It has become obvious that . . . the public . . . can easily fall prey to those critics of nuclear power who intentionally or unintentionally distort the truth about it." [1] In an editorial in *Physics Today* concerning the controversial effects of nuclear power, H. L. Davis claims that "Scientists in our midst have too often been less than fully responsible in statements and claims they have made in public." [2] This closing of the ranks of establishment scientists has only helped to convince the environmental protectionists that they face a tough battle with a scientific-industrial-governmental mafia which controls public policy on environmental matters.

Although it is natural to expect public controversy over issues that affect the quality of life and the public economy, it no longer appears possible for the scientist to stand aloof, contributing his expertise only when called upon. There are some who claim that scientific progress has so affected our society that all scientific work is inherently political, since it contributes to the advancement of governmental, industrial, or private group interests. Even if one cannot accept such an extreme view, he should recognize that it is widely held by many young scientists as well as activist groups in American society and is bound to affect the outcome of many environmental struggles.[3]

Beleaguered environmental scientists now recognize that they are the newest victims of the credibility gap that separates citizens from governmental, industrial, and private agencies. Statements by the Department of Defense, the Atomic Energy Commission, state public works departments, or local electric utilities are either disbelieved or greatly discounted by the average citizen. A scientist employed by any such agency must expect his work to be questioned by the public on grounds other than scientific ones. Rather than deploring this situation or papering it over by appealing for scientific unanimity, the scientific community should recognize its own fallibility and encourage open scientific debate and controversy. Nothing would so convince the public of a scientist's integrity as his biting the hand that feeds him.

Although the foregoing remarks might seem to be a digression, they touch upon issues that underlie much of what is discussed in this volume. If practical benefits are to result from the scientific studies discussed, means must be found to incorporate this information into the forming of public policy. The sponsorship of this symposium by the AAAS, and the publication of this volume, will help to make this possible.

References

1
G. T. Seaborg, "Looking ahead in nuclear power," *Mechanical Engineering, 91,* No. 8, 30–34 (August 1969).

2
H. L. Davis, "Clean air misunderstanding," *Physics Today, 23,* No. 5, 104 (May 1970).

3
W. Lockeretz, "Arrogance over clean air," *Science, 168,* 651 (1970), letter.

2

**Power Generation　　　　　Frederick E. Smith
and Human Ecology**

The technologies for alternate methods of waste-heat disposal are sufficiently developed so that any one of them can be used. The only substantive issue is the willingness of society to select something other than the method with the lowest direct cost. Unfortunately, in this social context the disposal of waste heat becomes a relatively small problem; it is but a part of the system of power production, which in turn is but a part of the system of power use and industrialization, which is but a part of the larger social system.

Our social system has evolved as a whole and in general terms has been shaped by the collective desires of the people. Within this system we have insisted on cheap, abundant power, and the power industries have succeeded in meeting our demands. Now that our skies and rivers are polluted, it is all too easy to point the finger at industry, and power production in particular, as the perpetrator of pollution. The fact is, of course, that pollution is the result of our concerted efforts for the last hundred years. The "benefits" resulting from this externalization of production costs have been pocketed by all of us.

Currently, the increasing rates of power use are only 20% dependent upon population increase; the remaining 80% comes from increased use per individual. This is not to imply that population growth is a minor problem — but it is certainly not the whole problem. Furthermore, the present environmental crisis is due only in part to increasing levels of pollution. The rest is

Editors' note. These comments were originally part of a panel discussion on waste heat at the AAAS symposium on Power Generation and Environmental Change. The discussion offered in this chapter, however, addresses broader questions of societal conflicts and values with respect to power generation, industrialization, and environmental quality and is appropriately included here.

due to an increasing perception of pollution resulting from demands for better environmental quality, which in turn is a byproduct of increasing levels of affluence. Here again, however, the rising levels of pollution are themselves a serious problem, with or without an increasing perception of pollution. As an aside, it can be noted that most of our forefathers were much too busy to learn how to swim in the clean rivers of this country, and it is only now, when we have the leisure to swim, that we cannot find clean water.

So long as our emphasis on cheap power persists, pollution will be a monumental problem. Except for token efforts with respect to a few obvious sources of pollution, the procedures for improving environmental quality are directly antithetical to the procedures for creating an abundance of power and materials. As the discussions in this volume demonstrate, industry itself is responsive to the demands of society and has throughout its development played an adaptive role that has been extraordinarily successful. A change in adaptation requires a change in societal demands. This can arise only from a reallocation of priorities. As several contributors to this volume have commented, "You have to give up something to maintain the environment."

Waste heat is also only one kind of environmental pollution, and pollution is only one kind of environmental deterioration. The total amount of rural and urban environmental degradation, which underlies many of today's social problems, has not to my knowledge been estimated by a competent, authoritative body. We are not talking about a few million dollars in cost here and a few million there, but a few million in many, many places. My own estimates are that the total cost of cleaning up the environment lies between fifty and several hundred billion dollars, a considerable fraction of the nation's economy. This suggests that we must consider the problem in terms of major tradeoffs between the cost of production and the cost of environmental quality.

If we choose as our fastest course of action the political area leading to legislation, two major paths can be contrasted. One is the program of subsidies to which Robert Jaske alludes (see Chapter 22), in which public funds are used to subsidize antipollution technology and industry, to reimburse industry for the cost of cleanliness, and in general to protect current market prices. The other is a program of regulations forcing industry to absorb the costs of cleanliness and the costs of associated research and development, ultimately forcing an increase in prices. The consumer pays in either case (there is no one else, since everyone is a consumer), either through higher taxes or through higher prices. Both methods will siphon off a portion of the economy that until now has gone into higher living standards and leisure time. The immediate effects of the two approaches seem similar.

The long-term effects, however, are decidedly different. Government subsidy involves a hidden cost. The individual consumer still pays his taxes, whether he consumes a little or a lot, and continues to pay low prices for power and goods. This will tend to result in a continued rapid growth in the use of energy and in the consumption of goods. Government regulation, by contrast, produces higher direct costs. The consumer is much more aware of where his income is going and will be more sensitive to additional purchases. In the long run, the rate of increase in the use of power and materials will slow down.

Expansion is deeply rooted in our industrial system, and power production is no exception. It exists in many of the rate systems applied to the consumer, whereby he pays less and less for additional increments of power. This encouragement for the increased use of power also encourages waste among the more affluent. Of greater concern, perhaps, is the correlation between power use and economic level. Under these graded price systems, the poor pay more per unit of power than the rich. Another way to state this is to say that the poor subsidize the rich in order to promote economic growth.

If the additional costs of clean production are added into this graded system, the greater burden will fall upon the small users of energy. It is this kind of expectation that has led those concerned with social issues to be less than enthusiastic over the attention being given to the environmental crisis.

Such conflicts can be resolved, however. To a considerable extent the present deterioration of the planet and the rising demand for environmental quality imply a shift toward a spaceship economy, one that conserves. In such an economy a pricing system in which large consumers pay lower rates is ridiculous. If the problem is one of conserving rather than using, the price system should be reversed. At present our society is not fully committed to zero-growth economics, but we are definitely moving away from free growth. Under these circumstances, perhaps, the cost rates for power use should be the same for all users and all rates of use. Then the burden of higher costs needed to produce a cleaner environment will be borne in proportion to use, and the poor will no longer subsidize the rich.

The policies to be adopted and the degree to which higher production costs should be accepted to produce a better environment are issues that should be decided by the public. They concern the management and future development of our society, issues that are certainly part of the public domain. Policies by edict from a government agency or by unilateral decisions within industries fail to involve the people as much as they should, unless they arise as responses to the expressed public will.

In closing I should like to note that the problems of environmental quality are the same in all developed nations, independent of their political structure. It does not seem likely, therefore, that a reorganization of structure will solve the problems. It is the basic philosophy of the people, expressed in the way social systems function, that seems to be at fault.

3

Future Needs for Power from Coal Wallace B. Behnke, Jr.

The environmental effects of electric power generation have in recent years become the subject of national attention and concern, and the AAAS Committee on Environmental Alteration must be commended for focusing attention on this critical issue.

I wish to discuss the future needs for power from coal. Although nuclear power is a major new source of energy, coal is today the dominant source of thermal energy for the electric power industry, and it will undoubtedly remain so over the next several decades.

But first a few words about my own company, Commonwealth Edison. By most measures, it is the third largest electric utility in the country and has been operating Dresden Nuclear Power Station for ten years. We burn more coal than anyone in the state of Illinois. We stand firmly for clean air, and we think nuclear power is a good way to reduce air pollution. We also stand firmly for diversity of energy sources and competition among the fuels we use. However, concern over air pollution has raised new doubts over the future of coal, and I have a few comments to make about this based on Commonwealth Edison's experience with fossil and nuclear generation.

Air pollution, which is almost entirely a by-product of the use of fossil fuels, is among the most serious of our environmental problems. It is a serious matter for all citizens in every city and in every developed country of the world. It is a by-product of our high-energy civilization. As Lee DuBridge stated to the Joint Committee on Atomic Energy:

The growth in use of energy is, of course, not an end in itself, but merely reflects the needs of a growing industrial nation. First of all, it reflects the increase in our population. . . . Perhaps more important in the growth of energy consumption is our rising standard of living and the greater mechanization of

The references for Chapter 3 are on page 19.

industry. . . . An abundant supply of low cost energy is the key ingredient in continuing to improve the quality of our total environment. . . . It seems to me, therefore, that it is absolutely crucial to recognize the fundamental importance of energy to our way of life. But at the same time, we must devote much greater effort than in the past to obtaining future supplies of energy with an absolute minimum degradation of the environment.[1]

At Commonwealth Edison we like to think of our role as extracting raw energy from its source, converting it to electricity — a more usable and an absolutely clean form — and then delivering it to the consumer for his use in improving the environment in which he lives, for example, through air conditioning, electric space heating, and better lighting. In our view, the environment can be better preserved by concentrating the conversion of raw energy at central electric-generating facilities equipped with the best waste-product control and disposal facilities that technology can provide, rather than by converting it as is done now in a multitude of small, inefficient fuel burning installations (such as home heating plants) dispersed within densely populated areas where air pollution is most severe.

Sharply increasing demands for electric power in this country are not a new experience. Almost from its inception eighty-five years ago, the electric power industry has approximately doubled its output every decade. By way of illustration, the industry produced 750 billion kilowatt hours in 1960 with about 175 million kilowatts of capacity. The production forecast for 1970 is 1.6 trillion kilowatt hours, with 340 million kilowatts of capacity. We see this growth rate continuing so that, in the year 2000, well over 10 trillion kilowatt hours will be produced. To give these figures added perspective, it will require about 2 billion kilowatts of generating capacity to produce the annual electric power requirements forecast for the turn of the century. Contrast this to today's generating capability of about 320 million kilowatts. About 80% of today's capacity is thermal capacity having a thirty-year economic life; we can safely assume that essentially all of these facilities will be replaced by the year 2000.

With power demands doubing every ten years, power producers must bear an enormous responsibility to provide this electricity, both where and when needed, within the area served, while maintaining reserve margins necessary to handle emergencies. They must anticipate future power needs, ordering machines and building facilities up to six years in advance. They must do what is reasonable to protect the environment. And, finally, they are obligated to do these things at a reasonable cost to the consumer.

What Are the Future Needs of Coal?
First let us consider the power industry's current commitment to coal. In 1970, the electric power industry is expected to consume 330 million tons of coal in generating facilities having an aggregate capacity of 260 million kilowatts. And even if our industry were to stop ordering coal-fired plants tomorrow, about 40% of this capacity would normally still be in service in 1990. This represents an enormous and largely unchangeable commitment to coal. By itself, this commitment is reason enough to push ahead as rapidly as possible on new technology for better limiting the air pollution from coal consumed in existing generating facilities.

Well-established technology for removing particulates from stack emissions is on hand and is being widely applied. Today's high-performance electrostatic precipitators should be adequate to meet air-quality requirements.

As for sulfur oxides, the picture is not encouraging. Stringent air-quality criteria have been issued by the Department of Health, Education, and Welfare, and are being translated into regulations. Moreover, public outcries about sulfur dioxide pollution are placing pressure on utilities to move faster in cutting down sulfur dioxide emissions. It is becoming increasingly difficult for utilities to meet the sulfur dioxide problem by the rural siting of generating facilities and the use of high stacks. Of course, there are substantial low-sulfur coal reserves. The eastern and midwestern reserves that are economically recoverable are largely committed to markets other than electric power production. Many of these coals have higher ash-fusion temperatures than the coals for which many existing boilers are designed to burn and therefore cannot be used unless major modifications to these boilers are made. And even if it were feasible to make these modifications, it would take a long time to do so because only a limited number of the boilers can be shut down at one time without jeopardizing the reliability of electric service. There are large uncommitted reserves of low-sulfur coal in the western states. But these are largely in undeveloped deposits, and getting this coal mined and shipped to utilities at great distances on a dependable basis is another matter.

At Commonwealth Edison we have tested about 30,000 tons of low-sulfur coal. One Wyoming coal plugged up the test boiler so that the generating unit had to be shut down and the rock-hard residue chipped out with jack hammers. In every case, the amount of particulate matter coming from our stacks doubled when we used these coals. No one wants to solve one problem while creating another. Although the results are discouraging, we have another 100,000 tons of western coal and will continue the tests.

What about stack-gas cleanup facilities? Despite more than thirty years of research and development, there is today no method that has passed the test of proved performance for removing diluted sulfur oxides from stack gases of large electricity-generating facilities. The Department of Health, Education, and Welfare pointed out in its 1968 annual report that more work needs to be done.[2] While there are a number of promising processes under development, these appear to be several years away from marketability on an assured performance basis. Of course, the utilities

3 Future Needs for Power from Coal

have to work with the technology available to them when they place orders for new facilities that often require up to six years to build. Even when we do get this technology, clearly there will be problems in fitting these processes to plants that have already been committed.

In the absence of technology for removing sulfur oxides from stack gases, some power companies such as Commonwealth Edison are switching to nuclear power, reducing their coal burn, and retiring older coal-fired plants located in highly urbanized areas on a programmed basis. We also substitute low-sulfur oil and natural gas for coal to the extent that these premium fuels are reliably available in the quantities required.

For the near future, we look for the premium fuels to displace coal to some extent in the existing power-production market. Since dwindling domestic supplies of gas are not adequate to fill the power industry's needs, more and more utilities will be looking to foreign oil and liquid natural gas.

Apart from the power industry's current commitment to coal, this fuel represents a huge and valuable resource that cannot be ignored. There are at least a trillion tons of coal in known, economically recoverable reserves in the United States alone. But growing public pressure to do something about air pollution is currently giving nuclear power an advantage, despite the recent nuclear construction delays, licensing difficulties, and rising prices. Both coal and nuclear power must assume increasingly greater roles in meeting the raw-energy requirements of the electric power industry. Both are vital natural resources that are essential to meeting future energy demands.

Most power forecasts indicate that in the year 2000 the bulk of generating capacity will be divided between coal and nuclear. If we reach the 2 billion kilowatt mark, nuclear and coal power could each account for about 900 million kilowatts of generation, while gas, oil, and hydroelectric plants will supply a relatively small portion of the capacity. This assumes that the fast-breeder nuclear reactor reaches commercial maturity in the mid-1980s and that we develop economic ways of dealing with the residuals from both energy sources so as to avoid unnecessary degradation of the environment. It seems to us, also, that it would be a serious mistake to abandon coal and foreclose to the public the benefits of competition between energy sources. Competition is an essential ingredient in our society because it encourages technological advances and holds prices down. With proper and reasonable standards and regulation, competition can be effective in serving the public interest and can help to clean up the environment faster.

In the period ahead, we look for the power producers' choice of fuels to turn more and more on how each can economically meet the environmental quality requirements American society will establish. If our turn-of-the-century forecast proves reliable, and coal gets half of the power-production market, the amount of coal-fired generating capacity in the United States will increase almost threefold from today's levels.

In my opinion, however, science and technology for dealing with air pollution from coal-fired plants are not keeping pace with the public demand for cleaner air. What is needed to keep coal in the running are some hard, practical solutions for limiting gaseous air contaminants from combustion. The coal producers should take more initiative in research and development. After all, it is their product and they should be taking the lead. The manufacturers, the government, and the utility users should also provide their share of support. What we need is more imagination and innovation and less confrontation. The public wants action, not words, and they are entitled to get it.

As a first order of business, at least several proved processes are needed for removing sulfur oxides from stack gases. Processes applicable to both new and existing plants must be developed as rapidly as possible — and these must be reliable and make economic sense — and they must not cause water pollution or increase emissions of other forms of air contaminants or degrade boiler availability.

Serious efforts should also focus now on the longer-term needs to limit nitrogen oxide and carbon dioxide emissions from the combustion process. They should be attacking the total problem of managing the residuals from the burning of coal. To minimize the economic penalty, they should seek practical ways of using by-products: the ash, sulfur, nitrogen oxides, metallics, heat, and even the carbon dioxide. The feasibility of integrated chemical extraction and power-producing complexes should be explored further. Such complexes would probably be large by today's standards and would be located in the coalfields. Valuable organic and inorganic constituents might be extracted and processed for market. The residual char could then be burned to produce electricity. Could nitrogen oxides be used to make fertilizer? Could the heated cooling waters be used to extend the growing season of crops in open fields or to raise commercial fish? Could carbon dioxide be used to stimulate plant growth? These are questions for the experts.

The ultimate coal complex will not be exactly like the one I have described, but I suspect that something similar could be built. It will take dedication and imagination, coupled with large amounts of time, money, and effort devoted to research and development. This could be coal's answer to the fast-breeder nuclear reactor. But unless there is such an effort to reduce the air contaminants from burning coal, the future outlook for coal will be clouded. What is needed is a massive, well-coordinated approach to assure that coal will remain a viable energy resource and a major competitive factor in our energy economy.

I am optimistic about the future. I am confident that the scientific and technical society that produced nuclear power and the manned space program can develop the means to protect the environment, preserve a viable competitive energy economy, and at the same time provide the great quantities of electric power needed for our ever expanding economy.

Discussion
Question from O. W. Stewart, Professor of Mechanical Engineering at the University of Kentucky.

Will electric power producers always remain at the mercy of combustion-equipment suppliers with regard to fuel used?
Answer by Wallace B. Behnke, Jr.

In selecting fuel-burning facilities for their generating stations, power producers are limited to the equipment that the manufacturers can supply. This is largely determined by the state of technology and competition. With competition from nuclear power and a growing public concern for environmental quality, we hope for accelerated development in fossil-fuel combustion technology. With respect to existing power plant facilities, it is necessary that the fuel be compatible with the facility in which it will be used. Also among the producers' considerations in selecting fuel are adequacy and reliability of supply, environment, and cost.

Question from Glenn L. Paulson, graduate fellow at the Rockefeller University and member of the New York Scientists' Committee for Public Information.

What barriers stand in the way of locating all pollution-producing power plants far from urban areas and with minimum impact on the environment of the distant areas?
Answer by Wallace B. Behnke, Jr.

Among the factors to be considered in siting large generating stations away from urban areas are the reliability of the transmission lines connecting them to load centers, availability of adequate cooling water, the adequacy and reliability of facilities for the transport of fuel, and the relative cost, as compared with alternative sites of generating facilities that are likewise compatible with the environment. With the development of large interconnected extra-high-voltage (EHV) transmission systems and gas-turbine peaking technology, it becomes increasingly feasible to locate future large generating complexes away from urban areas but no farther than necessary, in the interest of reliability.

References

1
Lee A. DuBridge, statement in *Environmental Effects of Producing Electric Power, Part I,* Hearings before the Joint Committee on Atomic Energy, 91st Congress, 1st session (Washington: U.S. Government Printing Office, 1969), pp. 3–30.

2
U.S. Department of Health, Education, and Welfare, *1968 Annual Report* (Washington: U.S. Government Printing Office, 1969).

II

Nuclear Power and Radionuclides in the Environment

4

Nuclear Power Reactors and the Radioactive Environment

Merril Eisenbud

The first nuclear reactor was demonstrated in December 1941, only thirty-six months after the discovery of nuclear fission. Under wartime pressure, reactor technology developed rapidly, and only one year later a 3.8-megawatt research reactor began operation at the Oak Ridge National Laboratory. It remained in service for more than twenty years, during which time it served as the chief source of radioisotopes for research in the United States and much of the Western world. Even more remarkably, the first of several reactors designed for plutonium production began operation at Hanford in 1944 at an initial power level of 250 megawatts. These units, at considerably higher power levels, also remained in service for more than twenty years without mishap.

At present, nearly 500 reactors have been constructed throughout the world. More than 100 power reactors are used on vessels of the U.S. Navy, and 21 privately owned nuclear power plants have been built in the United States. There are 82 additional civilian units that are currently under order. When constructed, they will produce about 70,000 megawatts of electricity, or about 14% of the expected national demand in 1975.[1]

The so-called light-water reactors are by far the most prevalent and are likely to be predominant well into the 1970s. This class of reactors includes two types, the boiling water reactors (BWR) and the pressurized water reactors (PWR). The fuel in both is usually slightly enriched uranium in the form of oxide pellets contained in stainless steel or zircaloy tubes. Water is used for both coolant and moderator. The tubes of fuel are arranged in

The references for Chapter 4 are on page 43.

bundles within which are interspersed control rods of neutron-absorbing materials such as hafnium, boron, or cadmium.

Uranium oxide has the desirable characteristic of trapping fission products within its crystal structure, thereby greatly reducing the probability of escape of fission products. This material is also advantageous because it avoids the hazards due to the pyrophoric nature of metallic uranium, which reacts exothermically with both water and air. It was this property of uranium that resulted in burnup of a part of a reactor core in Windscale, England, in 1956. As a result, radioactive iodine was distributed over the English countryside, necessitating confiscation of milk produced on nearby farms.[2] This is apparently the only reactor accident that has ever caused significant environmental contamination.

Power reactors are designed with a negative temperature coefficient of reactivity, a fundamental property of a reactor that contributes in a major way toward its safe operation. The effect of increased temperature in diminishing the reactivity of the reactor system is the result of several factors. When the fuel temperature is increased, the energy region in which nonfission capture takes place is broadened. If the water temperature should increase to the point where boiling occurs within the core, the steam voids would reduce the reactivity by providing less neutron moderation.

The pressurized water reactor (see Figure 4.1) includes a primary enclosure within which is the reactor core. It is a pressure vessel of several inches' thickness and is usually made of stainless steel (see Figure 4.2). Water is pumped through the core where it becomes heated by the nuclear reaction. Since the entire primary system is maintained under pressure, the water does not boil but passes through a heat exchanger (boiler), where it transfers its heat to a secondary system. The water of the secondary system boils within the heat exchanger and is pumped through a steam drier and then to the turbogenerator, from which the steam

4 Nuclear Power Reactors and the Radioactive Environment

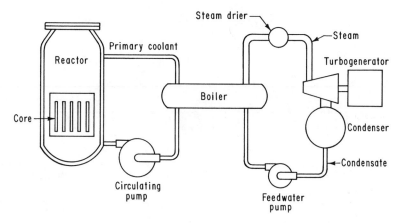

4.1
Schematic view of a pressurized water reactor (PWR) and the associated power generator.

tailings are condensed and returned to the boiler. Thus in the PWR the steam used to drive the turbogenerator does not pass through the reactor but receives its heat via the heat exchanger.

In the boiling water reactor (see Figure 4.3), the water passing through the reactor core is allowed to boil and the steam goes directly to the turbogenerator from which the steam tailings are condensed and returned to the reactor.

Radioactive Releases to the Environment

During normal reactor operation, traces of radioactive substances are released to the environment in both liquid and gaseous forms.

The water used to cool and moderate the reactor accumulates traces of corrosion products and other impurities, which become radioactive in passing through the reactor core. These activation products are mainly nuclides of chromium, cobalt, manganese, iron, and other metals and are usually present in higher concentrations than the fission products. Most of the fission products produced within the reactor core are retained in the uranium oxide

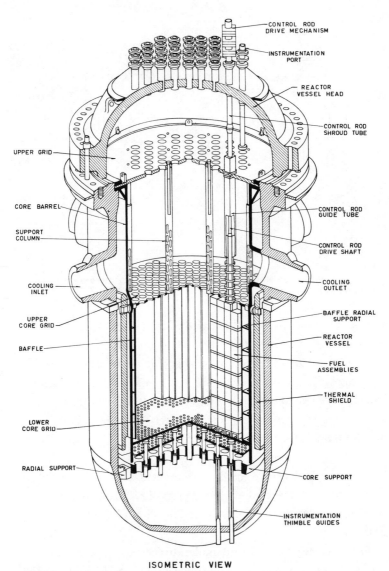

ISOMETRIC VIEW

4.2
Isometric cutaway view of the pressurized water reactor (PWR) No. 2 at the Indian Point Nuclear Power Plant, New York, showing the arrangement of the fuel assemblies, control rods, pressure vessel, and other components. This figure is taken from the *Directory of Nuclear Reactors,* Vol. VII, *Power Reactors* (Vienna: International Atomic Energy Agency, 1968), p. 55. It is reproduced with the permission of the IAEA.

4 Nuclear Power Reactors and the Radioactive Environment

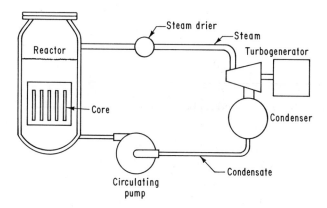

4.3
Schematic view of a boiling water reactor (BWR) and the associated power generator.

fuel, but certain of the more labile elements such as cesium, the halogens, the noble gases, and tritium do tend to diffuse from the fuel in very small amounts.

The buildup of radioactive contamination within the coolant can be controlled by continuous purification. The coolant is filtered to remove suspended solids and is then passed through cation and anion exchange resins and, if necessary, through a gas stripper from which dissolved gases can be removed.

The largest volume of liquid waste is normally generated in the course of routine maintenance or refueling operations. When the reactor system is serviced or when quantities of contaminated coolant must be disposed of, the liquids produced can be stored in waste-holding tanks to permit short-lived nuclides to decay. The wastes can then be passed through a gas stripper, after which the liquid can be decontaminated by filtration, cation-anion exchange, or evaporation. These processes result in concentration of the radioactive substances as a sludge or in solid form, and the waste materials can be placed in drums, mixed with concrete, and disposed of by shipment to AEC-approved storage sites.

Although largely decontaminated, the effluent from the decontamination process continues to contain traces of a number of radionuclides and is stored in holding tanks until it can be sampled and the radioactivity assayed. The radioactive concentration having been determined, the waste liquid can then be released slowly into the discharge of the condenser cooling system. Most power reactors are located on bodies of water to facilitate dissipation of heat from the condensers. Since the condenser coolant flow is apt to be several hundred thousand gallons per minute, this stream offers a convenient method of releasing the liquid waste diluted to the concentration permitted by the AEC license. The distribution of radionuclides in the liquid waste discharge from a typical light-water reactor is shown in Table 4.1.

The quantities of radioactive gases released from a reactor vary greatly depending on whether it is a pressurized water or boiling water reactor. The gaseous releases from the PWR are so insignificant as to be relatively less than the atmospheric dis-

Table 4.1
Typical Analysis of Radionuclides Released in Liquid Wastes from Water Reactors

Radionuclides	Release rate $\mu Ci/sec$
H-3 (tritium)	5×10^{-2}
Co-58	2×10^{-2}
Co-60	3×10^{-2}
Sr-89	8×10^{-3}
Sr-90	9×10^{-4}
I-131	1×10^{-3}
Cs-134	2×10^{-3}
Cs-137	6×10^{-3}
Ba-140	5×10^{-3}
Ce-144	2×10^{-4}

Note. μCi = microcuries; 1 curie = 3.7×10^{10} disintegrations per second. See the Glossary on page 412 for a more complete discussion of nuclear units.
Source. Bernard Kuhn et al., *Radiological Surveillance Studies at a Boiling Water Nuclear Power Reactor*, Bureau of Radiological Health (Washington: U.S. Public Health Service, 1969).

4 Nuclear Power Reactors and the Radioactive Environment   29

charges from fossil-fuel plants. The burning of coal or oil releases several radionuclides to the atmosphere, including radium-226, radium-228, uranium, thorium, and potassium-40. These are, of course, released in very small quantities, and there has been no suggestion that they represent a public health hazard; but the radioactive gaseous releases from pressurized water reactors have been shown to be even less than those from fossil-fuel plants, when due allowance is made for the relative potential hazards of the individual nuclides involved. (See Chapter 9.)

In the boiling water reactor, the steam passes directly from the reactor to the turbine and carries with it copious amounts of relatively short-lived noble gases as well as traces of other volatile fission products. In the typical boiling water reactor system the waste gases that pass through the turbine with the steam are stored for 30 minutes to allow the short-lived nuclides to decay, after which the radioactive gases are passed through filters before being released to a stack at a rate that complies with AEC standards. The composition of the radioactive mixture discharged from a typical boiling water reactor is shown in Table 4.2.

Table 4.2
Principal Radionuclides Released in Gaseous Emissions from Boiling Water Reactors

Radionuclides	Half-life	Release rate μCi/sec
^{m}Kr-85	4.4 h	3×10^2
Kr-85	10.7 yr	1×10^{-1}
Kr-87	76 min	7×10^2
Kr-88	2.8 h	5×10^2
^{m}Xe-133	2.3 day	1×10^1
Xe-133	5.3 day	3×10^2
Xe-135	9.1 h	8×10^2
Xe-138	17 min	2×10^3
H-3 (tritium)	12 yr	5×10^{-2}

Note. ^{m}Kr and ^{m}Xe refer to metastable states of the nuclides which decay by isomeric transition.
Source. See Table 4.1.

Limits of Permissible Release of Radioactive Wastes

Under federal law, protection of the public from effects of radioactivity is the responsibility of the Atomic Energy Commission (AEC), and all reactors are subject to licensing procedures that govern their design, construction, and operation. The reactor-licensing procedure is a complicated one that is spread over several years.

The AEC has relied on the National Council on Radiation Protection and Measurements (NCRP) and the International Commission on Radiological Protection (ICRP) to recommend the permissible dose for atomic energy workers and the public. The AEC has assumed for its part the role of translating the recommendations of these non-AEC expert groups into administrative language that lends itself to use by regulatory authorities.

The scientific bases for the ICRP and NCRP recommendations have been subject to independent review on a continuing basis by a number of independent national and international bodies. For example, since 1955 there has existed a United Nations Scientific Committee on the Effects of Atomic Radiation, for the purpose of reviewing and organizing the world's information on the effects of ionizing radiations. It is not the function of this committee to recommend levels of permissible exposure, but its compilations and analyses of the world's experimental and epidemiological information serve as a valuable aid in evaluating the ICRP recommendations.

Within the United States government, a Federal Radiation Council (FRC) was established by Presidential Order in 1959 to review all federal radiation standards. The FRC includes representatives from several departments of government, including Health, Education, and Welfare. Thus, although the AEC has responsibility for the public health aspects of its programs, the framework within which it operates is that provided by this complex of national and international organizations,

4 Nuclear Power Reactors and the Radioactive Environment

among which there has been total harmony of both aims and methods.

The AEC regulatory machinery includes another independent group established by congressional action, the Advisory Committee on Reactor Safeguards (ACRS). This committee, which consists of scientists and engineers from universities and industry, is charged with responsibility for reviewing all applications for AEC reactor licenses. The deliberations of the committee serve as an independent check on the parallel reviews given by the AEC staff.

The United States Public Health Service and the various states also play important roles. The Public Health Service maintains a staff that reviews all reactor license applications along with the AEC. The Bureau of Radiological Health of the Public Health Service also provides the states with financial and technical assistance so that they can monitor the air, water, and biota in the vicinity of nuclear reactors, with particular emphasis on vectors of human exposure.

According to AEC regulations, the maximum permissible per capita annual dose received by the public should not exceed 0.17 rem, and the maximally exposed individual should not receive more than 0.5 rem.* To put these figures into perspective, the whole body dose from natural sources of radioactivity in most parts of the world is about 0.1 rem per year; but there are wide deviations. Some inhabitants of areas of Brazil and India, where the natural radioactive levels are elevated by the presence of radioactive minerals in the soil, are exposed to as much as 100 times the average, or about 10 rem per year.[2]

The lung and skeleton are selectively exposed by natural radioactive sources over and above the dose received by the body as a whole. The added lung exposure is due to atmospheric radon,

---

* See the Glossary on page 412 for an explanation of the units of radiation exposure.

which can deliver a dose of as much as 1.3 rem per year to the more radiosensitive tissues of the lung. Doses as high as ten times this value have been reported in buildings made of materials with a relatively high radium content.

There are also places in the United States, as elsewhere, where the radium content of drinking water and food is elevated, resulting in abnormally high skeleton doses to people living in the area.

The ICRP and NCRP standards for protection against internal exposures are based on the assumption that the amount of radioactive substances accumulated within the body or in the organ that receives the highest dose (the critical organ) should be less than the amount that causes the permissible annual dose to be exceeded. These figures are then translated into maximum permissible amounts that can be inhaled or ingested, using a set of physiological parameters that describe the movement of each inhaled or ingested radionuclide to the critical organ.

The AEC regulations include limits on the permissible concentrations of radionuclides in air and water, and the regulations are frequently administered on the assumption that, if the maximum permissible concentration is not exceeded at the point of discharge to the environment, the dose to humans will not be exceeded anywhere beyond the site boundaries. The point of release in the case of a radioactive liquid effluent is the point at which the waste is discharged to the receiving body of water. As applied to gaseous effluent, the point of release is at the top of the stack, perhaps 500 feet from the ground. In most cases, this is a very conservative assumption, since dilution up to several orders of magnitude can take place beyond the point of release. However, it is also possible for biological concentration to take place, and when this occurs, the risk can be correspondingly increased. Thus it is known that iodine is concentrated by cow's milk, and many metals are concentrated in shellfish, sometimes ten thousandfold.

The AEC has in practice always placed upon the prospective licensee the responsibility of demonstrating that such concentration does not take place, although the AEC regulations have not been completely clear on this point until recently. Within the past few years, the AEC standards have been more specific. In the case of iodine-131, the maximum permissible concentration in air has been reduced by a factor of 700 to allow for the tendency of iodine to deposit on forage and eventually pass to cow's milk. Additionally, the regulations now state specifically that the licensee must demonstrate that accumulations in the food chain are not taking place.

The AEC also requires the licensee to conduct monitoring programs in the vicinity of the reactor. This provides information about the concentration of radioactive substances in air, water, and biota, including whatever food products may be grown in the vicinity. Thus the question of human safety is not left to conjecture but is based on actual measurement of samples collected from the environment. The states, assisted by the U.S. Public Health Service, undertake independent monitoring programs as well and serve to evaluate the monitoring results reported by the licensee.

Experience in the Atomic Energy Commission's Program
When the atomic energy program began more than twenty-seven years ago, the prospects must have seemed ominous in view of the deplorable safety record that had been accumulated by users of ionizing radiations prior to World War II. It is estimated that in the first forty years of this century only about two pounds of radium were extracted from the earth's crust — but this relatively small amount killed at least 200 people. Fortunately the uses and misuses of ionizing radiation prior to World War II were studied with great care by scientists and engineers from many nations, and the basic standards for safe industrial use of radioactive substances that became available in their earliest

forms just before World War II have served very well to protect the more than 200,000 atomic energy workers in the United States. Whereas about two pounds of radium used prior to 1940 had killed an estimated 200 people, the Manhattan District and the Atomic Energy Commission increased the amount of radioactive material available for industrial use many millionfold with an enviable safety record. In twenty-seven years, there has been a total of six deaths from nuclear accidents, all of which occurred in the course of experimental research or development.[3] In addition, there has been one death from a nuclear accident in a private industrial company. Among this large population of industrial workers, and over this relatively long period of time, I have been unable to find a single record of death or disease from the cumulative effects of exposure. Nor is there any indication that the incidence of neoplasms is higher among atomic energy workers than in the general population. This is a highly significant fact, since the population of atomic energy workers is a sizable sample of the total population, and atomic energy workers are allowed much higher exposures than the general population. The AEC standards limit the annual dose to the maximally exposed individual in the general population to one-tenth of the dose permitted for occupational exposure. The average dose to the public must not exceed one-thirtieth of the occupational dose.

To put the six accidental deaths into perspective, there was a total of 276 on-the-job accidental deaths from all causes, such as vehicular accidents, falls, and fires — these despite the fact that the overall record of industrial safety in the AEC program is one of the best in the country. Thus, while the record of radiation safety has not been perfect, potential risks of using large quantities of radioactive materials are being controlled better than the other occupational hazards with which we are all familiar.

The quantities of radioactive substances actually released to the environment by nuclear reactors are a small fraction of the permissible amount and thus far have not resulted in doses that come anywhere near 0.17 rem per year, even for residents in the immediate vicinity of the reactors. Excluding tritium, a total of about 50 curies of mixed fission and corrosion products were discharged to the adjacent waterways during 1968 by all civilian power reactors in the United States. In contrast, weapons testing deposited about 2 million curies of strontium-90 on the surface of the United States and many million curies of other radionuclides.[4]

Tables 4.3 and 4.4 summarize the gaseous and liquid releases from the operating civilian reactors during 1968. It can be seen that the amounts being discharged are but a fraction of that permitted by the AEC.

Tritium is released in much larger quantities from nuclear reactors, but its physical and biological properties are such that the permissible discharges are much greater than other nuclides with which we are concerned. None of the power reactors discharged more than 1% of the permissible amounts of tritium during 1968 (taken as a typical year).

The potential hazard from a given nuclide depends on the physical and chemical form in which it is released, the radioactive half-life of the nuclide, its decay scheme, and its behavior in biological systems. These differences are reflected in the quantities that can be ingested on a daily basis. Thus the permissible amount of soluble radium-226* in drinking water consumed by the general population is limited to 10^{-8} microcuries per milliliter (μCi/ml), whereas the permissible concentration of manganese-54 is 10^{-4} μCi/ml, a factor of ten thousand greater than the permissible radium concentration. The permissible concentration of tritium in drinking water is 3×10^{-3} μCi/ml, which is three hun-

---

* Radium-226, a naturally occurring substance, is one of the most hazardous of the radionuclides. There are about 5 million curies of this nuclide distributed more or less uniformly in the top of the earth's crust in the United States.

Table 4.3
Releases of Radioactivity from Power Reactors in Liquid Effluents, 1968

	Mixed fission and corrosion products			Tritium	
Facility	Curies released	Concentration limit (μCi/ml $\times$ 10^{-7})	Percent of permissible limit	Curies released	Percent of permissible limit
Big Rock	7.9	1.5	59.0	34.0	<1
Humboldt Bay	3.2	1.0	20.0	7.2	<1
Dresden 1	5.97	1.0	19.0	2.9	<1
Connecticut Yankee	3.8	1.0	7.3	1,735	<1
La Crosse	.074	1.0	3.0	Negative	—
San Onofre	1.5	1.0	2.5	2,353	<1
Saxton	.009	1.0	2.3	7.5	<1
Indian Point 1	34.6	35.0	2.0	787	<1
Elk River	.2	120.0	<1	8.2	<1
Peach Bottom	.00035	1.0	<1	Negative	—
Yankee	.009	1.0	<1	1,170	<1

Source. U.S. Atomic Energy Commission.

Table 4.4
Releases of Radioactivity from Power Reactors in Gaseous Effluents, 1968

	Noble and activation gases			Halogens and particulates		
Facility	Curies released	Curies permissible	Percentage of permissible	Curies released	Curies permissible	Percentage of permissible
Humboldt Bay	897,000	1,560,000	57	0.45	5.6	8
Elk River	648	19,000	3.4		0.1	<1
Dresden 1	240,000	22,000,000	1.1	.15	100	<1
Yankee	0.66	8,400	<1		0.03	<1
Big Rock	232,000	31,000,000	<1	.09	38	<1
Indian Point 1	55.2	1,600,000	<1		7	<1
Peach Bottom	108.5	12,600	<1		0.009	<1
Saxton	18.6	3,750	<1		10	<1
Connecticut Yankee	3.7	95,000	<1		0.2	<1
San Onofre	4.75	567,000	<1		0.8	<1
La Crosse	(<1)	480,000	<1		0.8	<1

Source. U.S. Atomic Energy Commission.

dred thousand times greater than the permissible concentration of radium-226. This is particularly important when considering the radioactive waste gases released to the atmosphere by boiling water reactors, the effluents of which contain biologically inert noble gases with short half-lives. The permissible concentration in air of a nuclide like xenon-135 is many orders of magnitude higher than for iodine-131.

Two nuclides in the effluents of nuclear reactors have attracted special long-range interest. These are krypton-85, which has a half-life of 10.5 years, and tritium (H-3) with a half-life of 12 years. Both are produced in copious amounts, and their properties are such that they do not concentrate in biological systems but can be expected, because of their relatively long half-lives, to accumulate in the environment.

It has been estimated that, with the anticipated growth of nuclear power, krypton-85 will accumulate in the atmosphere to the extent that by early in the twenty-first century it may double the natural radioactive background.[5] Accumulation of this nuclide in the earth's atmosphere is being documented, and should its concentration increase to the point where control seems desirable, it could be eliminated by freezing it from the gaseous wastes of power reactors or fuel-reprocessing plants. Being a noble gas, it cannot be eliminated from the waste streams by any known practical chemical process; but it is susceptible to removal at cryogenic temperatures.

Also, because tritium occurs in aqueous form, it cannot be eliminated from waste streams by any known practical method. At the present time, most of the tritium content of the hydrosphere is the result of rainout from nuclear weapons tests; the increment from reactor operations cannot be detected except within a short distance of the points of discharge. For the foreseeable future the tritium released by power reactors will not constitute a measureable incremental radiation exposure to the general population. The future quantities of tritium that will be produced by nuclear power plants can be estimated from projected requirements for nuclear power, and it can be shown that about 100 million curies will be produced throughout the world by the year 2000.[6] The worldwide inventory of naturally occuring tritium produced by cosmic rays is also about 100 million curies. It is estimated that, if uniformly distributed, the annual dose from reactor-produced tritium will be about one microrem per year by the year 2000. This dose is insignificant, even if one allows for the fact that uniform distribution is not likely to exist and that tritium may possibly be relatively more effective in producing genetic mutations than most other sources of exposure. The radiobiology of tritium is certainly deserving of special attention, and further research is required before the long-range implications of reactor-produced tritium can be evaluated fully.

Strontium-90, a nuclide that has received widespread public attention, is not released from reactors in significant amounts. Of the 50 curies of radionuclides released to the aquatic environment by all civilian power reactors in the United States in 1968, the strontium-90 content was much less than 0.1 curie. As noted earlier, this compares to 2 million curies of strontium-90 deposited in our country from weapons tests.

Discussion

Certain conclusions can be drawn concerning the impact of nuclear power reactors on the radioactive environment.

For more than a quarter of a century, the record has been a good one. The industry has achieved an admirable record of safety among its more than 200,000 employees, and the levels of environmental radioactivity have been subject to strict supervision in the vicinity of all AEC and private installations. The dose to the general population from civilian power reactors is in most cases not measurable against the normal background of natural radioactivity.

Although examination of twenty-seven years of experience

indicates that the AEC has been prudent in discharging the responsibilities that the U.S. Congress bestowed on it regarding health and safety, the fact remains that the public is not convinced that this is so. There remains a credibility gap that has not been closed after more than fifteen years of debate. A significant factor in the credibility gap is the unusual dual responsibility of the AEC for both development of civilian nuclear power and protection of the public health. Although it is readily apparent that the AEC has an excellent record of accomplishment in both areas and has retained a high degree of objectivity in facing its responsibilities for health and safety, the public is not fully convinced. For this reason it would be in the public interest to begin active consideration of the means by which the regulatory responsibilities of the AEC can be shared with some other agency of government. Only in this way can we hope to assure the public that the present apparent conflict of missions is not operating to its detriment. But a transfer of regulatory responsibility cannot be accomplished easily. The AEC has well-developed regulatory machinery of a type that does not exist in any other branch of government. Whereas in theory it would be possible to transfer this organization *in toto* to another agency, this would not be wise because governmental interagency transfers are frequently disruptive of morale and working efficiency.

As a compromise, the Public Health Service should be given a more prominent role in the regulatory program. The Public Health Service, rather than the AEC, should promulgate the numerical standards of permissible exposure. The AEC, with its highly developed capability to evaluate reactor designs, should continue to consider applications for new reactors and should continue to monitor construction and operation to assure compliance with the terms of the license. But the Public Health Service, in its traditional collaborative relationships with the states, should undertake the responsibility of effluent monitoring and ecological surveillance. By sharing its present statutory regulatory authority with the Public Health Service in this way, the credibility gap that now exists between the AEC and many segments of the public might be closed.

The levels of radiation exposure permitted by the AEC are those recommended for worldwide use by the International Commission on Radiological Protection. Though neither the ICRP nor the AEC is sacrosanct, considerable weight must be given to the fact that the ponderous procedures of these organizations have produced a set of regulations that are workable and have successfully protected the public health for more than a quarter of a century. The present system of AEC regulation, which puts major emphasis on the maximum permissible concentrations of radionuclides in air and drinking water, should be changed in favor of specifying the maximum permissible daily intake from all sources. This is the method used by the Federal Radiation Council and is preferable because it automatically takes into consideration such factors as multiple sources of exposure and ecological factors. Although this change is desirable, there is no great urgency about the matter since the way in which AEC administers the present regulations accomplishes the desired objectives.

There are now nearly 500 reactors throughout the world. The features being incorporated into the designs of new reactors take advantage of all that has been learned in twenty-seven years. If the record has been good up to now, it should be better in the future.

The nature of many of the past activities of the AEC was such as to necessitate the release of quantities of radioactive materials to the environment on a scale vastly greater than is required in the civilian power industry. Relatively large releases to the environment at major research and production centers such as Hanford, Oak Ridge, and Savannah River have provided excellent opportunities for radioecological studies. Fallout from weapons testing has provided similar opportunities. Nearly three decades of research have produced much of the basic in-

4 Nuclear Power Reactors and the Radioactive Environment

formation with which one can predict the behavior and effects of radioactive substances in the environment.

These being the facts, how does one explain the substantial opposition to nuclear power plants, particularly since the alternative to nuclear power often requires the combustion of fossil fuels and release of noxious gases and dusts into the atmosphere? At worst, one can argue that radioactive pollution, even at the AEC's permissible levels, may cause finite health effects that are too small to measure by existing investigative techniques but are effects that nevertheless may exist. In contrast, combustion of fossil fuels involves environmental and human effects that are readily indentifiable. Sulfur dioxide in combination with particulates in the atmosphere has clearly been associated with disability and death because of its effects on the cardiovascular and respiratory systems. This acid gas also causes materials to deteriorate and is known to harm plant life. The ecological effects of SO_2 have not been studied, and hence our knowledge of its effects is confined to obvious pathology in certain species of higher plants. Moreover, we know that combustion of fossil fuels releases CO_2 to the atmosphere in copious amounts and that the effluents contain known carcinogens as well as many trace substances of a toxic nature. Why the clamor about nuclear plants? Why won't the public accept a small risk in exchange for the greater known risks from burning coal or oil?

The answer is not simple. To some extent the attitude of the public has its roots in a quite justifiable concern about nuclear war. The general public cannot readily break the mental connection between nuclear reactors and Hiroshima. The AEC has surely suffered from being associated in the public mind with the Bomb, and that association has obscured its activity as a developer of the peaceful uses of the fission process. The public loses sight of the fact that the AEC has conducted an admirable basic research program and that atomic energy has played an important role in many of the great medical developments of our time. Nor do they know that the atomic energy industry was the first industry to accept its responsibility for environmental protection and that its standards of responsibility are widely applauded by health officials. The AEC's concern for the environment, developed under the watchful eyes of the Joint Committee on Atomic Energy of the U.S. Congress, antedated by twenty years the development of similar attitudes in other industries.

Many scientists, both inside and outside the AEC program, have played a useful role in relation to the public understanding of nuclear energy. In addition to writing and lecturing about the subject in a responsible way, they have provided constructive criticism that has been invaluable to government and industry in charting a socially responsible course for the civilian atomic energy program.

Regrettably, not all scientists have been constructive or, I might add, fair in their approach to the public. A few individuals in the scientific community have capitalized on the general ability of scientists to catch the ear of the public but have then proceeded to disregard the disciplined, well-ordered approach to issues which the scientific method requires and which is after all the reason why scientists have achieved and deserved a high degree of credibility in their relations with the public.

One of the most effective means of accrediting the views of a scientist is the traditional system of preparing well-documented papers for review by peers and publication in scientific journals. For all its defects, this system has worked well, but unfortunately it is being bypassed by many scientists in favor of the newspaper, the public meeting, or the popular magazine. These media are excellent for the airing of views on major issues, but public understanding of the issues might be better served if some way could be found to disassociate the scientist from special privilege when he speaks outside his field of specialization and when he bypasses the time-tested methods developed by scientists to assure the credibility of scientific information. Perhaps then the public will be able to form its judgments on the basis of information on reactor performance and safeguards within a context defined by the pollution potentialities of the alternatives.

References

1
U.S. Atomic Energy Commission, *Annual Report of the U.S. Atomic Energy Commission* (Washington: U.S. Government Printing Office, 1970).

2
M. Eisenbud, *Environmental Radioactivity* (New York: McGraw-Hill, 1963).

3
U.S. Atomic Energy Commission, *Operational Accidents on Radiation Exposure Experience* (Washington: U.S. AEC, December 1968).

4
M. Eisenbud, "Radionuclides in the environment," presented at the symposium: Diagnosis and Treatment of Deposited Radionuclides (Richland, Washington: May 15–17, 1967), Excerpta Medica Foundation.

5
J. R. Coleman and R. Liberace, "Nuclear power production and estimated krypton-85 levels," *Radiological Health Data and Reports*, 7, No. 11, 615–621 (November 1966).

6
D. G. Jacobs, *Sources of Tritium and Its Behavior upon Release to the Environment* (Oak Ridge, Tennessee: U.S. AEC, Division of Technical Information, December 1968), also available as TID-24635, from Clearinghouse for Federal Scientific and Technical Information, Springfield, Virginia.

5

Nuclear Reactors and the Public Health and Safety Arthur Tamplin

Over the past several months, I have been involved in a number of conferences throughout the country held to present the pros and cons concerning the installation of nuclear power reactors. No one at these conferences was strictly against the installation of the reactors. Instead, the concern was to insure that power reactors were installed and operated safely, that they were sited appropriately, and that everyone was properly apprised of the potential biological effects of the radioactivity released to the environment by the reactor. The purpose of this chapter is to discuss the biological consequences of the radioactivity released by reactors.

The Maximum Permissible Concentrations
In normal day-to-day operations, nuclear power plants are permitted by law to release radioactive atoms to the environment in gaseous and liquid discharges. There are two regulations applied to these releases. One represents the primary standard, which is the dosage that can be delivered to an individual or to the population at large. The other regulation consists of a set of secondary standards. These are the maximum permissible concentrations (MPCs) of various radionuclides in air and water that can be released outside of the restricted area of the plant. One should be derivable from the other, but the secondary standards, the maximum permissible concentrations listed in Title 10 of the Code of Federal Regulations,[1] do not permit this. They do not take into account the biological concentration mechanisms that actually take place in the environment between the release of

The references for Chapter 5 are on pages 58–60.

5 Nuclear Reactors and the Public Health and Safety

the activity by the plant and the eventual consumption of contaminated foods by man.

A group at the Lawrence Radiation Laboratory has developed a practical technique to estimate the dosage to man from each radionuclide that would be released to the environment in nuclear activities. Our approach takes into account the physical and biological concentrating mechanisms that intervene between the release of radionuclides to the environment, their subsequent uptake in the food chains, and the consumption of these foods by man. The approach treats all routes of entry to man: cow's milk, direct ingestion of surface-contaminated foods — soil to plant to man — and fresh and marine aquatic food chains.

The results of calculations related to the release of radionuclides to a freshwater system are presented in this chapter. More complete and detailed analyses are available in a series of reports we have prepared on this subject,[2-6] all of which are available on request. One of these reports[5] is a handbook that contains all of the input data necessary for the calculations, as well as a set of numbers that allows easy conversion of radioactivity releases into dosage to man. These conversions have been derived, using the other basic data tabulated in the handbook.

I should now like to show how this approach can be used to determine the yearly release that could be allowed from a reactor without exceeding the present guideline dosage to the average individual of a population or to the most critical individual within a population. I will illustrate this by making use of a hypothetical river. This stretch of river is some 200 kilometers long, 300 meters wide, 10 meters deep, and has a volume of 600 million cubic meters. I also assume that the water in the river is replaced each day; that is, the water flows at about 5 miles per hour. I also assume that there are some 5 grams per square centimeter of bottom material in equilibrium with the water, which is about the first inch of bottom sediment. This hypothetical river is not an unreasonable river in terms of many rivers for which reactors are planned. Finally, I assume that the population exists totally on a diet of aquatic origin derived from this river. This means that the limiting releases, therefore, would apply to the most critical individual in the population, that person who is obtaining a large fraction of his diet from this river. Table 5.1 shows the results of these calculations.

The first column in the table shows the curies per year (Ci/yr) of a few radionuclides that would lead to a bone dosage of 0.5 rad/yr (the allowable dosage for individuals). The radionuclides listed are the ones that would be among the most critical radionuclides; that is, the ones that would have the most limiting release to the river. This column shows that if 20 Ci/yr of antimony-125 were released to the river, the dosage to bone would reach the value of 0.5 rad. The same dosage would be produced by 100 Ci/yr of strontium-90, 300 Ci/yr of ruthenium-106, and 600 Ci/yr of cesium-137. The next column in Table 5.1 lists the picocuries per liter (pCi/liter) that would be achieved in the water at equilibrium. In other words, this is the concentration in water that would result in a dosage of 0.5 rad/yr. As you can see, these numbers are quite small, varying from 0.1 pCi/liter for antimony-125 to 3.0 pCi/liter for cesium-137. The next column in the table

Table 5.1
Amount of Radionuclides Resulting in 0.5 rad/yr to Bone from Hypothetical River

Radionuclide	Quantity released to river (Ci/yr)	Equilibrium concentration in river water (pCi/liter)	MPC in water, federal regulation (pCi/liter)
Sb-125	20	0.1	1×10^4
Sr-90	100	0.4	3×10^2
Ru-106	300	1.0	1×10^4
Cs-137	600	3.0	2×10^4

Note. 1 curie = 3.7×10^{10} disintegrations per second; 1 pCi = 10^{-12} curie; Ci/yr = curies per year; MPC = maximum permissible concentration. See the Glossary on page 412 for a more complete explanation of the units of radiation exposure.

5 Nuclear Reactors and the Public Health and Safety 47

lists the MPC in water from Title 10 of the Code of Federal Regulations.[1] The MPCs in water are in every case considerably higher than the concentrations that would actually result in a dosage of 0.5 rad/yr to this most critical individual. As a matter of fact, even if we assume that average individuals are receiving only 1% of their diet from the river, we still find that most of these MPCs are extremely high compared to the concentrations listed in column 2, with the exception of strontium-90. In this case, the values in column 2 would have to be multiplied by 100. The reason for this difference is that the MPCs tabulated in Title 10 of the Code of Federal Regulations do not take into account the biological concentrating mechanisms that intervene between the release of radionuclides by the reactor and their eventual accumulation in biological organisms that form the food base for man.

At the end of one year's operation, a 1,000-megawatt plant would contain some 3×10^6 curies of cesium-137. The 600 Ci listed in Table 5.1 represent only 2 parts in 10,000 of the cesium-137 inventory of the reactor. In other words, the nuclear industry can release only very small fractions of its radionuclide inventory into an aquatic system without producing appreciable dosages in at least some members of the population. The other important point in this respect is that for a river on which a number of reactors and one or more fuel-reprocessing plants are planned, one has to consider all of this nuclear activity in setting the release rates from any one of the plants because they will add together in producing the dosage.

In principle, there is every reason for believing that the nuclear industry can release much less radioactivity than the limits listed in Table 5.1. In practice, however, we find that, at least in 1967, a fuel-reprocessing plant in New York State was exceeding these limits in its associated aquatic system.[7,8] The average concentration reported for strontium-90 in the associated stream water was 24 pCi/liter. Antimony-125 was measured in the waste lagoon but not in stream water or fish. As a scientist who is concerned about informing the public concerning the safety of the operation of the nuclear power industry in this country, I am disturbed by the reprocessing plant. The monitoring of reactors and reprocessing plants should include all of the critical isotopes that are indicated by a careful analysis of the existing biological and physical data. Before we can truly attest to the dosage that will be delivered to the population as a result of nuclear power activities, it is essential that a comprehensive study be made. This study must take into account both physical and biological concentrating mechanisms and be based upon quantitative data on each and every radionuclide in the inventory of the entire nuclear power industry in each ecological region of the nation. It would then be possible to determine the limits of radioactive discharges that should be applied to the plants. It seems quite certain at this moment that these limits will indicate that it is absolutely essential for the industry to be designed and operated so as to approach, and to approach quite closely, the absolute containment of the radioactivity within the reactors and fuel-reprocessing facilities.

The "Primary" Radiation Exposure Guidelines

The primary standard that should be applied to all aspects of atomic energy is the most critical factor related to the day-to-day operations of the nuclear power industry. This primary standard is the dosage that should be allowed to be given to the average individual in the population as a result of the operation of this industry. Once we have arrived at an acceptable dosage to the population at large or to the most critical individual within the population, we can then proceed from this number and calculate backward to determine the amount of radioactivity that can be released by the reactors into the environment without exceeding this primary dosage limitation. The present guidelines set forth by the Federal Radiation Council and adopted by the Atomic Energy Commission in Title 10 of the Code of Federal Regulations

5 Nuclear Reactors and the Public Health and Safety

are that the average dosage deliverable to the population at large is 0.17 rem per year.[1] The other standard is that the limiting dosage deliverable to an individual is 0.5 rem per year. It is also stated in the federal guidelines that the dosage should be kept as far below this limit as is practicable. It is important to recognize that the nuclear power industry, acting in what I would say is a responsible manner, has built extra safety into the reactors. Its design objectives have been to limit the exposure of the public to something less than 0.01 rem per year — that is, less than 0.1 of the allowable guidelines.

But the point is that there is no solid basis for believing that the 0.17 rem per year guideline is safe. In fact, scientific evidence suggests that if the United States population were exposed to this dosage, the result could be some 17,000 additional deaths from cancer each year.[9,10] This is more deaths each year than have resulted from the worst year of the Vietnam War. The 17,000 deaths are estimated even without any consideration of genetic effects, which would be expected to cause even more damage. The power industry must realize that its design objectives do not necessarily contain a margin of safety. In fact, it is essential that the dosage to the public at large be kept at or below the design objectives.

Radiation and Cancer

I should now like to discuss the estimate of the number of cancer deaths that would be produced by the 0.17 rem per year guideline proposed by the Federal Radiation Council. We arrived at this estimate by examining the available data on the induction of human cancers by radiation. The basic data that went into this analysis were derived from the Atomic Bomb Casualty Commission studies of the Japanese who were irradiated as a result of the atomic bombs dropped on Hiroshima and Nagasaki. In addition to this, we also used the data on the medical irradiation for the disease rheumatoid spondylitis, the data concerned with the medical irradiation of infants for thymic enlargement, and the data derived from studies of women who were given routine x-ray pelvimetry during pregnancy.

Table 5.2 summarizes the data derived from the irradiation of children and adults. This table shows the dosage of radiation that would double the spontaneous incidence of these cancers and also the percent increase in the incidence rate per rad of radiation delivered.

There is a remarkable similarity between the doubling dose and the percent increase in the incidence rate per rad for all cancers. For such a widely divergent array of organ systems, already including data for nearly all the major forms of human cancer, it is indeed amazing that there is such a small range in the estimated doubling dose. The only number that is different, and that indicates an even higher susceptibility to radiation induction of cancer, is for thyroid cancer induction in persons under twenty years of age. We shall see subsequently that this is not at all surprising or inconsistent, for the data to be presented

Table 5.2
Best Estimates of Doubling Dose of Radiation for Human Cancers and the increase in Incidence Rate per Rad of Exposure

Human cancers by type or location	Doubling dose	% increase in incidence rate per rad
Leukemia	30–60 rads	1.6–3.3%
Thyroid cancer		
in adults	100	1.0
in young persons	5–10	10–20
Lung cancer	∼175	0.6
Breast cancer	∼100	1.0
Stomach cancer	∼230	0.4
Pancreas cancer	∼125	0.8
Bone cancer	∼40	2.5
Lymphatic and other hematopoetic organs	∼70	1.4
Carcinomatosis of miscellaneous origin	∼60	1.7

Note. Doubling dose refers to that dosage which would double the natural, spontaneous incidence.

suggest a very high susceptibility to irradiation of embryos in utero.

Stewart and his co-workers originally,[11] and MacMahon[12,13] and Stewart and Kneale[14] more recently, have presented evidence that implicates in utero radiation of embryos, carried out for diagnostic purposes in the mother, in the development of subsequent leukemia plus other cancers in the first ten years of the life of the child. The general estimate of the amount of radiation delivered in such diagnostic procedures is 2–3 rads to the developing fetus. From the Stewart and Kneale data, we have the estimates of the increase of cancers for several organ sites for various forms of cancer. These are shown in Table 5.3. Also included in the table are the data from MacMahon.

Again we see a very remarkable similarity in the data with respect to the increase in cancer for a wide variety of sites. These data, therefore, suggest that 2–3 rads delivered to the developing

Table 5.3
Estimates of the Increase of Cancers

Type of cancer	% of radiation-induced increase over spontaneous incidence
From the Stewart and Kneale data	
Leukemia	50%
Lymphosarcoma	50
Cerebral tumors	50
Neuroblastoma	50
Wilms' Tumor	60
Other cancers	50
From the MacMahon data	
Leukemia	50
Central nervous system tumors	60
Other cancers	40

Source. A. Stewart and G. W. Kneale, "Changes in the cancer risk associated with obstetric radiography," *Lancet*, *1*, 104–107 (1968); B. MacMahon, "Prenatal x-ray exposure and childhood cancer," *J. Natl. Cancer Inst.*, *38*, 1173–1191 (1962); B. MacMahon and G. B. Hutchinson, "Pre-natal x-ray and childhood cancer: a review," *Acta Unio Intern. Contra Cancrum*, *20*, 1172–1174 (1964).

fetus in utero will result in a 50% increase in the incidence rate of various cancers, and this leads to an estimate of 4–6 rads as a doubling dose for childhood leukemia plus cancers due to diagnostic irradiation in utero. In view of the widely diverse forms of human cancers plus leukemia showing striking similarity in the risk of radiation induction, it does not appear at all rash to propose some fundamental laws of cancer induction by radiation in humans.

Law I.

All forms of cancer, in all probability, can be increased by ionizing radiation, and the correct way to describe the phenomenon is either in terms of the dose required to double the spontaneous incidence rate of each cancer or, alternatively, as an increase in the incidence rate of such cancers per rad of exposure.

Law II.

All forms of cancer show closely similar doubling doses and closely similar increases in the incidence rate per rad.

Law III.

Youthful subjects require less radiation to increase the incidence rate by a specified fraction than do adults. The developing fetus requires a still lower amount of radiation.

Based on these laws and the extensive data laready in hand and already described, the following assignments in Table 5.4 appear reasonable for all forms of cancer.

Considering all the data that bear upon the proposed risk estimates, we consider these general laws as having a better experimental base than many laws of physics, chemistry, or biology when they were first proposed. Furthermore, we would estimate that the numbers probably underestimate the risk. For purposes of setting radiation tolerance guidelines, one might even be advised to use lower doubling doses than those estimated here.

In estimating the increase in cancer deaths per year, we used a risk estimate of 1% increase in cancer incidence rate per rad of radiation delivered. This was not the most restrictive number

Table 5.4
Assigned Doubling Dose and Incidence Rates

Adults	~100 rads as the doubling dose ~1% increase in incidence rate per year per rad of exposure
Youthful subjects (less than 20 years of age)	Between 5 and 100 rads as the doubling dose Between 1% and 20% increase in incidence rate per year per rad of exposure
Infants in utero	~6 rads as the doubling dose ~17% increase in incidence rate per year per rad of exposure

we could use, because the in utero radiation and the childhood radiation suggest the risk per rad would be much higher for those individuals. The death estimate was then made on the basis that in 30 years at 0.17 rem per year an individual would receive an integrated dosage of 5 rad. This would result in a 5% increase in the subsequent cancer death rate. Since we have ignored the in utero and childhood radiation and have also not used the animal data suggesting that all causes of deaths rather than cancer alone would be increased by radiation, the 17,000 additional deaths per year we estimated represent a minimum estimate of the effects of this radiation.

On the other hand, there are two hypotheses generally pro posed which would argue that our risk estimate is an overestimate of the risk at this level of radiation exposure. The first argument concerns the possibility of a threshold dose below which no harm is accrued to man as a result. The major support for this threshold dosage comes from Evans's analysis of radium-exposed persons.[15] My colleague J. W. Gofman and I have prepared a separate report on the subject.[16] The human radium exposure data do not represent any substantive evidence that a threshold dose does exist. Additionally, the data on the irradiation of infants in utero show effects at dosages of 2–3 rads. Even more important, these data indicate that each rad may even be ten times more effective in inducing cancer at these extremely low doses than is the case for each rad at the higher dosages. The threshold concept is without any solid scientific basis. It is still possible that a threshold effect does exist for man, but to use the hope that such a threshold may exist in setting guidelines for the exposure of the population seems unwarranted.

The second argument against our risk estimates is that delivering radiation slowly over a period of years, as would be the case in the peaceful application of atomic energy, may be less harmful with respect to cancer induction than the same dose delivered rapidly. This argument suggests that a repair mechanism exists. When the radiation is delivered at a low dosage rate, a mechanism is able to repair previously induced damage so that the net effect of a total dosage is reduced. There is actually some evidence from studies on mice suggesting that a repair of genetic damage is possible. But there is no evidence for such an effect on cancer and leukemia induction by radiation in man. Furthermore, the radium painters received their radiation slowly over a period of years, and it would appear that any protection that this might provide is not enough to alter our conclusions. I will present some additional evidence below which suggests that one should treat this repair mechanism extremely cautiously with respect to man.

The Genetic Consequences of Radiation

The more recent data on the biological effects of radiation generally tend to demonstrate that any original optimism about the effects was wrong. In this respect I cite the extreme radiosensitivity of the developing fetus in utero. To a considerable extent the existing guidelines were based upon the effects of radiation on adults; the data are now demonstrating that the developing fetus is from 10 to 100 times more sensitive than the adult. In addition, the data suggest that leukemia is not the most sensitive form of cancer with respect to radiation, but that indeed all cancers are equally likely to be induced. Further, the data show that other cancers occurring in the population more frequently than leukemia are induced by radiation in proportion to their occurrence rate. Finally, the data now coming in from biological experimentation suggest that the most radiosensitive portion of the biological system was overlooked in setting the original standards. This part of the biological system is the chromosomes, that is, the packages of genes. Radiation can affect genetic material in several ways. One way, which has been given the most attention with respect to the genetic and somatic effects of radiation, is the production of point mutation. By this process, irradiation changes the structure or the composition of a single gene, that is, a single hereditary unit. On the other hand, it is now abundantly clear that radiation can also affect the chromosomes. The radiation is able to alter or remove from the genetic material not a single gene but a large number of genes. The developing evidence on the effects of radiation on chromosomes suggests that this process may represent the major mechanism through which radiation produces its damage both somatically and genetically. In other words, a whole new mechanism for the potential biological effects of radiation is now evolving: a mechanism that may represent a far greater susceptibility of man than any previous mechanism proposed.

I think it extremely important to point out that scientists are not omniscient. Even though we have a considerable body of information at our disposal, we can never be certain that we have made all the pertinent observations necessary to determine the outcome of a particular series of events. We must always keep in mind that we do not necessarily have all the significant facts before us when we are asked to make recommendations whether something that is planned will adversely affect man or his environment.

Subsequent to the establishment of the exposure guidelines, a whole new body of experimental data concerning the radiosensitivity of chromosomes has evolved. Recent results reported by a group from The Johns Hopkins University demonstrate quite well the importance of the new body of data with respect to the biological effects of radiation.[17] The John Hopkins data indicate that between 1 and 2 rads delivered in the first 30 weeks of in utero life will produce severe genetic damage to the fetal germ cells, and in this case it appears to be chromosomal damage. As a result of this damage, 50% of the female conceptuses of women who were themselves irradiated as fetuses will be killed. This is a very startling observation, and other confirmations of this observation are highly desirable. It is an effect whose magnitude far exceeds anything that had previously been predicted concerning the genetic effects of radiation. Note also that this radiation was delivered some 20 years earlier. It seems that little repair if any has occurred.

One of the existing dogmas concerning the genetic effects of radiation is that only sex-linked mutations would be expected to contribute to deaths in the F_1 generation (the children born to irradiated parents). Sex-linked lethal mutations would express themselves by the deaths of male fetuses. As a consequence of sex-linked lethal mutations, one would expect to find the male-to-female ratio to be depressed in the live births. However, the Johns Hopkins study, rather than showing a depression in the male-to-female ratio, indicated that the male-to-female ratio increased. Not only did it increase, it doubled. In other words, twice as many boys were born as girls from mothers who were irradiated as fetuses at dosages in the range of 1 to 2 rads.

The FRC guidelines for the population at large would allow 0.1 rad to be delivered in the first 30 weeks of in utero life. Considering the Johns Hopkins data, this could result in the deaths of some 2.5% of the female conceptuses. It would appear only reasonable to assume that an equal fraction of the conceptuses would be affected by the nonlethal, but nevertheless severe, life-shortening chromosomal changes. All I am suggesting here is that this is most likely not an all-or-none phenomenon; nonlethal changes in the chromosomes could also be expected to occur. It would not seem unreasonable to assume that in addition to the 2.5% of the female conceptuses that are killed, something on the order of 1% of the surviving live-born infants might have severe or significant chromosomal changes. Actually, the Johns Hopkins data suggest differences in the live-born infants. Considering that each year in the United States we have some 4 million live births, this would suggest that as many as 40,000 severely affected children might be born if all mothers were, as fetuses, previously exposed to in utero irradiation.

    The above analysis is somewhat speculative. Nevertheless, I think the rather dramatic divergence from the accepted genetic dogma shown by the Johns Hopkins study suggests that we should examine very critically this entire area as it relates to the allowable dosage to the population. If nothing else, it points out that when applying the existing 0.5 rem per year dosage standard deliverable to an individual, rather than the 0.17 rem per year dosage deliverable to the population at large, one should apply this to the most significant or the most critical of human beings, namely, the developing fetus in utero. During the first 30 weeks of intrauterine life the 0.5 rem guideline would suggest that the fetus could receive 0.3 rad. In considering the effects observed in the Johns Hopkins study, this would suggest that approximately 7.5% of the female conceptuses of those mothers exposed to the limiting dosage deliverable to an individual of 0.5 rem per year would be killed, and certainly one would then have to assume that a significant number of the live births, male and female, might carry with them severe chromosome changes that could result in significant life shortening and severe disease in future life. In other words, here is a potential genetic effect of radiation that does not require the multiplication by 200 million people in order to demonstrate its significance.

In summary, when all of the data are considered and proper respect is given to the promotion of public health and safety, there would seem to be no other course than to reduce the existing radiation exposure guidelines. It would seem that these guidelines should be reduced by at least a factor of 10.

References

1
Code of Federal Regulations (10 CFR 20), Title 10 — Atomic Energy. Part 20 — Standards for Protection against Radiation (Washington: Office of the Federal Register, National Archives and Record Service, General Services Administration, revised as of January 1, 1969), pp. 134–158.

2
A. R. Tamplin, *Prediction of the Maximum Dosage to Man from the Fallout of Nuclear Devices,* "I. Estimation of the maximum contamination of agricultural land," Report UCRL-50163, Part I (Livermore: University of California Radiation Laboratory, 1967).

3
Y. C. Ng and S. E. Thompson, *Prediction of the Maximum Dosage to Man from the Fallout of Nuclear Devices,* "II. Estimation of the maximum dose from internal emitters," Report UCRL-50163, Part II (Livermore: University of California Radiation Laboratory, 1968).

4
C. A. Burton and M. W. Pratt, *Prediction of the Maximum Dosage to Man from the Fallout of Nuclear Devices,* "III. Biological guidelines for device design," Report UCRL-50163, Part III, rev. 1 (Livermore: University of California Radiation Laboratory, 1968).

5
Y. C. Ng, C. A. Burton, S. E. Thompson, R. K. Tandy, H. K. Kretner, and M. W. Pratt, *Prediction of the Maximum Dosage to Man from the Fallout of Nuclear Devices,* "IV. Handbook for estimating the maximum internal dose to man from radionuclides released to the biosphere," Report UCRL-50163, Part IV (Livermore: University of California Radiation Laboratory, 1968).

6
A. R. Tamplin, H. L. Fisher, and W. H. Chapman, *Prediction of the Maximum Dosage to Man from the Fallout of Nuclear Devices,* "V. Estimation of the maximum dose from internal emitters in aquatic food supply," Report UCRL-50163, Part V (Livermore: University of California Radiation Laboratory, 1968).

7
N. I. Sax, P. C. Lemon, A. H. Benton, and J. J. Gabay, "Radio-ecological surveillance of the waterways around a nuclear fuels reprocessing plant," *Radiological Health Data and Reports, 10,* No. 7, 289–296 (1969).

8
W. J. Kelleher, "Environmental surveillance around a nuclear fuel reprocessing installation, 1965–1967," *Radiological Health Data and Reports, 10,* No. 8, 329–339 (1969).

9
J. W. Gofman and A. R. Tamplin, "Low dose radiation and cancer," *IEEE Transactions on Nuclear Science, NS-17,* No. 1, 1–9 (Part I, February 1970).

10
J. W. Gofman and A. R. Tamplin, "Federal Radiation Council Guidelines for radiation exposure of the population-at-large — protection or disaster?" testimony presented before the Subcommittee on Air and Water Pollution, Committee on Public Works, U.S. Senate, 91st Congress, 1st session, November 18, 1969.

11
A. Stewart, J. Webb, and D. Hewitt, "A survey of childhood malignancies," *Brit. Med. J., 1,* 1495–1508 (1958).

12
B. MacMahon, "Pre-natal x-ray exposure and childhood cancer," *J. Nat. Cancer Inst., 38,* 1173–1191 (1962).

13
B. MacMahon and G. B. Hutchinson, "Prenatal x-ray and childhood cancer: a review," *Acta Unio Intern. Contra Cancrum, 20,* 1172–1174 (1964).

14
A. Stewart and G. W. Kneale, "Changes in the cancer risk associated with obstetric radiography," *Lancet, 1,* 104–107 (1968).

15
R. D. Evans, in *Radiation Exposure of Uranium Miners,* hearings before the Joint Committee on Atomic Energy, 90th Congress, 1st session (Washington: U.S. Government Printing Office, 1967).

16
J. W. Gofman and A. R. Tamplin, "Studies of radium-exposed humans: the fallacy underlying a major 'Foundation of NCRP, ICRP, and AEC guidelines for radiation exposure to the population-at-large,' " supplement to testimony presented before the Subcommittee on Air and Water Pollution, Committee on Public Works, U.S. 91st Congress, 1st session, November 18, 1969.

17
M. B. Meyer, T. Merz, and E. L. Diamond, "Investigation of the effects of prenatal x-ray exposure of human oogonia and oocytes as measured by later reproductive performance," *Am. J. Epidemiol., 89,* 619–635 (1969).

6

**Other Views on
Public Health and Safety**

I.
Introduction　　　　　　　　　The Editors

The generation of electrical power from nuclear energy has been the subject of increasing controversy. The power industry and the U.S. Atomic Energy Commission have been criticized for polluting the visual environment with large structures and transmission lines, the water environment with excessive amounts of waste heat, the air, water, earth, and biological environment with radioactive discharges, and man himself with radioactive poisons. Criticism regarding the visual and thermal aspects are justifiably directed to other types of power plants, as well as nuclear, and also to other grandly conceived engineering projects. The most heated and at times emotional interchanges, however, have concerned the effects of artificial introduction of additional amounts of radioactive substances into the world around us.

From the naïve enthusiasm of a decade ago about unlimited possibilities for exploiting the atom, many people seem to have swung toward the view that we have a scourge upon us. It can be argued that the opposition to nuclear power and radioactivity releases is merely symptomatic of a societal dissatisfaction with man's use and exploitation of the environment. Nuclear energy is well known to everyone, the result of aggressive promotion for many years, while the effects of radiation on living systems are somewhat mysterious and insidious and associated with dread diseases. There are ample tools by which an organized protest movement can "get" the nuclear establishment. The nuclear power industry and the AEC are required to bear the thrust of an attack that might more logically be directed at other, more careless exploiters of the environment who have remained less identifiable.

The power industry has been traditionally secretive about its plans for expansion and future development. This is less true of the AEC, however, although the question is being raised whether a single organization can be both developer and regulator of a new technology. Consequently, many responses to the environmental questions have been taken to be somewhat less than candid. Reconciliation of the different viewpoints becomes increasingly difficult when the opposing sides keep themselves apart by accusation or imperturbability. The complex technical nature of the questions involved does not simplify the controversy.

The material in this chapter is primarily a review and discussion of the material presented in Chapter 5, which has proved particularly controversial. Since the forum offered by the Symposium on Power Generation and Environmental Change did not permit adequate time for exploration of many questions, the editors of these proceedings invited others whose views are different or opposite to comment upon the subject matter of Chapter 5. The author of the chapter was not given the opportunity to respond to this criticism in these pages.

The editors encourage all readers to study the material in Chapters 5, 6, 7, 8, and 9 very carefully in the formation of their own opinions and judgments.

II.
Comments on "Nuclear Reactors and the Public Health and Safety"

Charles W. Edington, John R. Laughnan, George E. Stapleton, and Daniel W. Wilson

We have been invited by the editors of this book to comment on the chapter by Arthur Tamplin entitled "Nuclear Reactors and the Public Health and Safety." In accord with Tamplin, we feel that reactors should be sited properly and that their effluents should be carefully controlled. Further, we are in complete agreement that the best available information on the movement of radionuclides and the potential biological effects of radiation should be used to accomplish these objectives. Publication of such information has been common practice, and for many years, interpretation of the published data has been carried out by competent scientists in the development of national and international radiation protection standards for exposure of workers as well as the general public. It is our opinion, however, that Tamplin has provided neither an entirely accurate nor a sufficiently comprehensive analysis of the potential biological consequences of radioactivity released by reactors. Although we do not intend to analyze each of his statements, we have given attention to the three major areas discussed by Tamplin.

In his first section, of "The Maximum Permissible Concentrations," Tamplin argues that release limits are set too high because, as a consequence of reconcentration in the human food chain, releases of radioactivity at MPC levels could lead to body burdens and radiation doses higher than 0.5 rem per year. It is

Editors' note. This paper reflects the scientific opinions of the individual authors and is not to be construed in any way as an official response of the U.S. Atomic Energy Commission. It was received by the editors on May 1, 1970. The references for this section of Chapter 6 are on pages 72–74.

true that reconcentration of some radionuclides can occur, but it should be recognized that current regulations reflect this possibility and call for additional limitations where significant public exposure might occur as a result of reconcentration. For example, the current maximum allowable release limit for radioiodine gas and radioactive particulates takes into account the reconcentration of radioiodine and airborne particulate radioactivity via the food chain. The regulations also provide that additional limitations may be imposed as deemed appropriate. In addition, the regulations impose monitoring procedures that foster the release of radionuclides at concentrations well below the set limits. These regulations are set down in "Standards for Protection Against Radiation," in the Code of Federal Regulations.[1]

Some comments are in order concerning Tamplin's method [2] of arriving at concentration factors, as well as on the use of his model. He determines concentration factors for a specific radionuclide in the food chain leading to man by calculating the concentration ratio of the naturally occurring element in man and in the environmental substrate. This is a highly unrealistic model because it neglects the overall dynamics of the system, including atmospheric, geologic, and hydrologic transport phenomena, all the fractional contributions in the diet via the food-chain pathways, and the complete dynamics of nutrient flux in the ecosystem. Tamplin's compilation of food-chain concentration factors is recognized as a significant contribution to an understanding of the fate of radioactivity in the biosphere, but in seeking a solution to the problem it must be regarded as a starting point, not a perfected analysis.

In the third section, "Radiation and Cancer," we note from Tamplin's citations that his information is drawn from the same sources that have been analyzed previously by the national and international bodies of scientists responsible for establishing the guidelines questioned in his chapter. But Tamplin's use of the doubling-dose concept for radiation-induced cancer and his generalization that "all forms of cancer show closely similar doubling doses and closely similar increases in the incidence rate per rad," to estimate the excessively large number of deaths from cancer expected from an exposure of 0.17 rem per year, would appear to be unwarranted on the basis of the available scientific data. In fact the ICRP Task Group[3] recently pointed out:

In radiological protection the radiation dose required to double the natural cancer incidence is sometimes used in assessing acceptable risks from somatic exposure by analogy with the concept of doubling dose used in assessing the genetic risks from exposure of the gonads. This concept of doubling dose for somatic hazards is a specific example of the misuse of the ratio of cancer rates. The natural incidence of stomach cancer in men or women in five different countries varies between 65 and 706 per million living (Segi and Kurihara, 1963, cited by Dolphin and Eve, 1968)[4] so that for a fixed risk per rad the doubling dose varies more than ten-fold and will induce between 65 to 706 additional cases of stomach cancer per million persons depending on the particular population to which attention happens to be drawn. Superficially, the "doubling dose for cancer" may appear a reasonable concept because the overall incidence of all forms of cancer taken together happens to be roughly similar in many different countries. However, there are complex reasons for this and where acceptable risks and individual varieties of cancer are concerned, the only reasonable parameter to use is the actual number of cases induced by the exposure under consideration.

With respect to leukemia induction, Lewis[5] used three of the four sources of information cited in Tamplin's chapter and concluded that there is a probability of 1–2 cases of induced leukemia per million persons at risk per year per rem. In addition, Upton[6] has compared the data for leukemia induction in the Japanese survivors with those available for the individuals irradiated for ankylosing spondylitis. Both sources of data suggest a linear relationship between annual incidence per million person-years at risk and dose to the bone marrow. The "doubling dose" for the Japanese population is close to 100 rads while that for the spondylitics approximates 300 rads. Although these two estimates are not widely discrepant, considering the characteristics of the data from which they are drawn, they are not in accord with the

30–60 rad doubling dose shown in Table 5.2 of Tamplin's chapter.

In Table 5.2 estimates of incidence of thyroid cancer resulting from irradiation of young persons are included. A doubling dose of 5–10 rads is indicated. It is generally agreed that such estimates are hazardous because no individual survey includes sufficient cases to permit an accurate definition of the dose-incidence relationship. For example, the pooled data of Beach and Dolphin[7] suggest that a dose of 300 rads to the thyroid will double the incidence of malignant thyroid tumors in children. More recent data of Hempelmann[8] on a single large population of irradiated children indicate an increase of 2.5 cases per million persons per rad per year, a risk very close to that reported for leukemia in adults. Moreover, since Tamplin deals in particular with estimates of mortality, it should be pointed out that Hempelmann[8] and others[3] have emphasized the nonfatal character of thyroid cancer. It is also well documented that less than half of these thyroid tumors are cancers. On the basis of these data it is difficult to determine how the "best estimates" were arrived at by Tamplin. The extreme variations in tumor incidence from different geographic locations even within the United States make it difficult if not impossible to pool data.

The best consensus[3] of investigators working in the area of somatic effects of radiation is that the fetus is more radiosensitive than the adult. "Collation of ten separate investigations of antenatal radiography showed that none of them was out of line with the general conclusion that this form of foetal irradiation increased the overall risk of death from malignant disease between 0–10 years of age by about 40 percent." The mass of data in this case does accord with that presented in Tamplin's Table 5.3. Tamplin, however, fails to point out that, while the relative risk per unit dose is greater for in utero irradiation, the relative dose that would be accumulated during this sensitive stage is small as compared with what would be accumulated in postnatal

6 Other Views on Public Health and Safety

life. Both these factors must be considered where continuous low-level irradiation is concerned.

To arrive at the large number of additional deaths per year estimated by Tamplin, we assume that the quoted risk would have to apply to the entire population of the United States and that this population is irradiated in a homogeneously contaminated biosphere at a level that would result in an average dose of 0.17 rem per year. Earlier in the chapter Tamplin states that reactors are designed in such a way that an equivalence of less than 0.01 rem is released to the environment. If the developers, acting in a "responsible manner," as Tamplin indicates, regulate the reactor releases to less than 5% of the maximum permissible level, it is difficult to understand why the author is pessimistic about the future.

In the section on cancer Tamplin also raises the question of the effects of low dose rates or "delivering radiation slowly over a period of years" rather than rapidly. The radium dial painters are cited as an example of a situation in which there is no evidence for repair of damage leading to cancer. A cursory examination of the radiobiological literature would reveal that repair of damage produced by high ion-density radiation such as that emitted by radium would not be expected. On the other hand, evidence for repair or a decrease of damage produced by low ion-density radiations, such as x rays or gamma rays, is well documented. As an example, it is relevant to consider the large, carefully conducted animal experiments of Upton.[9] The data show that the radiation-induced incidence of myeloid leukemia in mice is increased about 100% per 100 rads of gamma rays when the exposure was made at 80 rads per minute. It is interesting to note that this compares favorably with Tamplin's estimate for man. But when the exposure was made continuously, at about 1 rad per day, no increase in leukemia over the spontaneous incidence was detected at total doses as high as 150 rads. In a companion experiment involving densely ionizing radiation (fast neutrons), no dose-rate dependence was found for the induction of leukemia. There are also numerous animal data[10] that demonstrate dose-rate dependence for life-span shortening. These late somatic effects represent accumulated damage of all types, including carcinogenic damage.

There are no comparable human data, but with the mass of presently available data on mammalian cells, including cultured cells of humans, it is not unreasonable to propose that intracellular repair of precarcinogenic damage occurs.

Later in his chapter Tamplin discusses again the radiosensitivity of in utero exposed populations from two points of view. (1) The low doses received (2–3 rads) suggest no threshold dose for cancer induction in the fetus. This has also been discussed at length by other authors, and there seems to be agreement. (2) The same data from low-dose irradiated fetal populations are then invoked again to suggest that "each rad may even be ten times more effective in inducing cancer at these extremely low doses than is the case for each rad at the higher dosages." The data used for a comparison of the relative radiosensitivity of pre- and postnatally exposed individuals are for the most part from low-dose exposed fetuses and high-dose exposed adults. One must accept these data to show either the greater sensitivity of the fetus as compared with the adult or the greater effectiveness of low doses as compared with high doses — but not both. The documentation provided by Tamplin does not allow us to decide which conclusion he has chosen.

In the last section of the chapter Tamplin presents his views on the genetic consequences of radiation. Here he asserts that "the most radiosensitive portion of the biological system was overlooked in setting the original standards. This part of the biological system is the chromosomes, that is, the packages of genes." He states that "a whole new mechanism for the potential biological effects of radiation is now evolving." Tamplin's comments apparently are based, in part, on his interpretation of the

6 Other Views on Public Health and Safety

results of an epidemiological study[11] carried out by Meyer, Merz, and Diamond, a team of researchers at The Johns Hopkins University. This investigation was designed to determine the effects of exposure to low doses of x rays, used in pelvimetry, on the reproductive characteristics of females irradiated in utero and on children born to them.

Tamplin's interpretations and extrapolations go far beyond the data in the Meyer, Merz, and Diamond publication and, in fact, are basically inconsistent with the interpretations of the authors themselves. For example, Tamplin states: "The Johns Hopkins data indicate that between 1 and 2 rads delivered in the first 30 weeks of in utero life will produce severe genetic damage to the fetal germ cells, and in this case it appears to be chromosomal damage. As a result of this damage, 50% of the female conceptuses of women who were themselves irradiated as fetuses will be killed." At another point Tamplin extrapolates to the general population as follows: "The FRC guidelines for the population at large would allow 0.1 rad to be delivered in the first 30 weeks of in utero life. Considering the Johns Hopkins data, this could result in the deaths of some 2.5% of the female conceptuses." These remarks are drawn from data in the Johns Hopkins paper on sex ratios among 490 children born to females who years before had been exposed in utero to x rays used in pelvimetry on their mothers. Although it was not stated by Tamplin, these data indicated that there was no significant difference in sex ratio of offspring born to mothers in the exposed and control groups. However, in analyzing the sex-ratio data as a function of the time during fetal life when the irradiation had occurred, Meyer, Merz, and Diamond[11] arbitrarily chose three periods of exposure: 0–29 weeks, 30–36 weeks, and 37 weeks to term. When this analysis was made, the sex ratio among children who had been exposed in the intermediate and late periods did not differ significantly from controls. Only 55 children (11% of the total) were born to females who had been exposed in the early period

(0–29 weeks). Among these children there were 37 males and 18 females, a sex ratio that represents a statistically significant deviation from the control sample. It is this deviant sex ratio in a sample of 55 children that Tamplin has chosen to regard as the result of death of half of the female conceptuses due to chromosome damage and typical of the entire population were it to receive equivalent exposures during the first 29 weeks of life. He also infers from these data that 0.1 rad delivered in the first 30 weeks of in utero life might lead to severe defects in 1% (40,000 per year) of live-born infants through the induction of nonlethal but severe life-shortening chromosomal changes. This is a misinterpretation and misuse of data.

The Johns Hopkins paper carried no data on chromosome damage, and no chromosome studies are reported. Moreover, no data are available to determine whether the deviant sex ratio is due to the selective loss of female conceptuses, to preferential survival of male offspring, or to sampling error. In fact, there is an indication that overall birth rates among mothers who were in the early exposed group are enhanced when compared to those for the two other groups. Moreover, the numbers of neonatal and fetal deaths and of ascertained abortions were similar in the exposed and unexposed groups. These observations are not consistent with a loss-by-death mechanism. It should be pointed out also that to our knowledge there is no genetic interpretation that can explain the deviation in sex ratio observed by Meyer, Merz, and Diamond in the 0–29 week in utero exposure period. There are three ways that chromosomes might be affected by radiation, but none of them would explain an increase in male births. For example, it is possible that dominant lethal changes might be induced; however, these would affect both sexes equally. Sex-linked recessive lethals or loss of an X chromosome would tend to give rise to a decrease in male progeny, an effect opposite to that observed. Another possibility, which the Johns Hopkins group has proposed[11] as a hypothetical basis for further analysis, is that irradiation of the fetus in the early period might induce nondisjunction of X chromosomes in oocytes (immature female reproductive cells), which might then lead to excess male births among the offspring. We note, however, that nondisjunction of the X chromosomes in oocytes is expected to produce a relative deficiency rather than an enhancement of male births. Apparently, in concluding for death through chromosome damage, Tamplin relied on this suggestion made by the authors despite its hypothetical nature. In addition, the authors suggested an alternate hypothetical scheme that Tamplin does not discuss. We read: "An alternative though unknown mechanism to explain the altered sex ratio in progeny of irradiated females might be that a hormonal change is induced by irradiation of the fetal F_1 ovaries. This might act in a way leading to preferential survival of male offspring."

Tamplin has asserted that "the most radiosensitive portion of the biological system was overlooked in setting the original standards" and that "a whole new mechanism for the potential biological effect of radiation is now evolving." Geneticists, on the other hand, have known about the mutagenic effects of radiation since Muller's pioneering work with x rays in 1927,[12] for which he received the Nobel Prize in 1946, and in the intervening years a vast scientific literature (for example, see References 13, 14, 15, 16) on the effects of ionizing and other forms of radiation on mutations and chromosomal aberrations in a variety of organisms, including man, has been accumulated. Many of these investigations, and others concerned with fundamental genetic phenomena, have been conducted in an effort to gain a better understanding of the mechanisms involved in the production of gene and chromosome changes and of the hazards connected with the genetic effects of radiation. As a consequence, geneticists, as well as other groups that have been convened to consider genetic effects of radiation, have pointed out that there is no dose below which genetic effects are not produced. Thus, in contradistinction to Tamplin's arguments, we contend that chromosomal changes have been studied for many years and have been considered in setting standards.

References

1
Code of Federal Regulations (10 CFR 20), Title 10 — Atomic Energy, Part 20 Standards for Protection against Radiation (Washington: Office of the Federal Register, National Archives and Record Service, General Services Administration, revised as of January 1, 1969), pp. 134–158.

2
Y. C. Ng, C. A. Burton, S. E. Thompson, R. K. Tandy, H. K. Kretner, and M. W. Pratt, *Prediction of the Maximum Dosage to Man from the Fallout of Nuclear Devices,* "IV. Handbook for estimating the maximum internal dose to man from radionuclides released to the biosphere," Report UCRL-50163, Part IV (Livermore: University of California Radiation Laboratory, 1968).

3
International Commission on Radiological Protection Publication 14, *Radiosensitivity and Spatial Distribution of Dose* (New York, London: Pergamon Press, 1969).

4
G. W. Dolphin and I. S. Eve, "Some aspects of the radiological protection and dosimetry of the gastrointestinal tract," in *Gastrointestinal Radiation Injury,* M. S. Sullivan, ed. (Amsterdam: Excerpta Medica Foundation, 1968), pp. 465–474.

5
E. B. Lewis, "Leukemia and ionizing radiation," *Science, 125,* 965–972 (1957).

6
A. C. Upton, "Radiation and carcinogenesis," *Ann. da Acad. Brasiliera de Ciencias, 39* (Supplement), 129 (1967).

7
S. A. Beach and G. W. Dolphin, "A study of the relationship between x-ray dose delivered to the thyroids of children and the subsequent development of malignant tumors," *Physics in Med. and Biol., 6,* 583 (1962).
8
L. H. Hempelmann, J. W. Pifer, G. J. Burke, R. Terry, and W. R. Ames, "Neoplasms in persons treated with x-rays in infancy for thymic enlargement. A report of the third follow-up survey," *J. Nat. Cancer Inst., 38,* 317 (1967).
9
A. C. Upton, "Comparative aspects of carcinogenesis by ionizing radiation," *Nat. Cancer Inst. Monograph, 14,* 221 (1964).
10
C. D. Van Cleave, *Late Somatic Effects of Ionizing Radiation* (Oak Ridge, Tennessee: U.S. Atomic Energy Commission, Division of Technical Information, 1968), pp. 23-72.
11
M. B. Meyer, T. Merz, and E. L. Diamond, "Investigation of the effects of prenatal x-ray exposure of human oogonia and oocytes as measured by later reproductive performance," *Am. J. Epidemiol., 89,* 619-635 (1969).
12
H. J. Muller, "Artificial transmutation of the gene," *Science, 66,* 84-87 (1927).
13
D. E. Lea, *Action of Radiations on Living Cells,* 2nd ed. (Cambridge: Cambridge University Press, 1955).
14
Report of the United Nations Scientific Committee on the Effects of Atomic Radiation, U.N. General Assembly document, 21st session, Supplement No. 14 (A/6314) (New York: United Nations, 1966).

15
Report of the United Nations Scientific Committee on the Effects of Atomic Radiation, U.N. General Assembly document, 24th session, Supplement No. 13 (A/7613) (New York: United Nations, 1969).
16
Alexander Hollaender, ed., *Radiation Biology,* Vol. 1, Part 2 (New York: McGraw-Hill, 1954).

III.
Comments on "Nuclear Reactors and the Public Health and Safety"

Merril Eisenbud

These comments are concerned with Arthur Tamplin's methods of analyzing the exposure of surrounding populations to radioactive substances discharged from a nuclear power reactor into a hypothetical river.

The basic method used by Tamplin was developed by him in connection with his studies of the environmental behavior of radioactive debris from nuclear weapons. This approach is not applicable to nuclear reactors for a variety of reasons. The fuel in contemporary power reactors is uranium oxide clad in zircaloy or stainless steel, a system that greatly inhibits the release of fission products. Moreover, there are great differences in the rates at which the nuclides of the various elements diffuse from the fuel. For this reason three of the four isotopes in Tamplin's Table 5.1 are among the most minor constituents of reactor effluents. These are antimony-125, strontium-90, and ruthenium-106. While strontium-90 is produced abundantly in the fission process, its escape-rate coefficient from uranium oxide fuel is 3 orders of magnitude lower than more labile nuclides such as those of cesium, iodine, and the noble gases.

The radiochemistry of reactor effluents is well documented. The nuclides in reactor effluents are primarily activation products, such as manganese-54 and cobalt-60, rather than fission products. The data of Tamplin's Table 5.1 thus are a radionuclide mixture whose composition is unrelated to those of actual reactor effluents.

So much for the qualitative aspects of Tamplin's assumptions.

Editors' note. This paper was received by the editors on June 2, 1970.

The quantitative assumptions seem likewise unreasonable. Tamplin has calculated that 100 curies of strontium-90 per year discharged into the hypothetical river from his hypothetical reactor would produce a dose of 0.5 rad per year to the bones of residents subsisting on a diet of aquatic food from the river. That 100 curies per year of strontium-90 could be discharged from a single reactor is an unreasonable assumption. As a matter of fact, all of the civilian reactors in the United States during 1968 discharged only about 50 curies of total radionuclides excluding tritium (H-3). The strontium-90 content of the liquid effluents from the nuclear reactors was probably less than 0.1 curie in the entire country. In his calculations, Tamplin assumed that 600 curies per year of cesium-137 were discharged into the hypothetical river. In the United States in 1968 the total amount of cesium-137 discharged from all operating reactors was of the order of 1 curie.

Tamplin has calculated that the hypothetical river would contain a strontium-90 concentration of 0.4 picocurie per liter (pCi/liter) from the reactor discharge and that this in turn, by ecological processes, would result in a dose of 0.5 rad per year to an individual who obtained a large fraction of his diet from the river. This is inconsistent with actual observation. Fallout from testing nuclear weapons has deposited about 2 million curies of strontium-90 on the United States, and this has been reflected by concentrations of strontium-90 in surface waters that for some years were of the order of 1 pCi/liter, higher than the 0.4 pCi/liter which Tamplin suggests would produce a bone dose of 0.5 rad per year. Although there has been extensive monitoring of foods throughout the United States for more than a decade, I am unaware of any data suggesting that 1 pCi/liter of strontium-90 in surface waters could result in a bone dose of 0.5 rad per year. In fact there is abundant evidence that this is not the case. The dose to the general population from strontium-90 deposited by weapons tests has been shown by the United Nations Scientific

6 Other Views on Public Health and Safety

Committee on Effects of Atomic Radiation to be very much less than 0.5 rad per year. In fact, the "infinity" or lifetime dose was estimated by UNSCEAR to be about 0.064 rad, with the dose deriving from contaminated milk and grain rather than fish. This may stem in part from the fact that strontium-90 goes mainly to the bones of fish rather than to the edible portions.

Tamplin continues to insist that the tables of maximum permissible concentration in Title 10 of the Code of Federal Regulations do not take into account the biological concentrating mechanisms that "intervene between the release of radionuclides by the reactor and their eventual accumulation in biological organisms that form the food base for man." However, the maximum permissible concentration for iodine-131 in air does include a factor of 700 to allow for concentrating mechanisms, and the text of the regulations specifically require that the licensee demonstrate, by environmental monitoring, that biological concentration does not take place, regardless of what nuclides are involved.

Tamplin notes that a 1,000-megawatt plant would contain about 3 million curies of cesium-137 at the end of one year of operation. He then notes that 600 curies, his assumed discharge rate, represents only 2 parts in 10^4 of the cesium-137 inventory. From this he concludes that the nuclear industry can release only very small fractions of its radionuclide inventory into an aquatic system. Cesium is one of the more labile fission products, but the release to the environment of this nuclide was about 10^{-6} of the reactor inventory, a factor of 100 lower than Tamplin's assumed release rate. The release of strontium-90 was less than 10^{-10} of the reactor inventories due to the less labile properties of this nuclide. His assumption is thus unreasonable by a factor of about 10,000 in the case of cesium-137 and by a very much larger amount in the case of the other nuclides he has used.

IV.
Low-Level Radiation and Health of the Public

Shields Warren

Three-quarters of a century have given us considerable experience with artificial ionizing radiation; the last quarter-century, with rapidly increasing numbers, kinds, and sizes of radiation sources, has seen great interest, much experimental work as to effects, and some controversy as to possible effects on the environment and on man. All of us now have absorbed at least a few atoms of strontium-90 and cesium-137.

If we had applied the same rigorous standards for the maintenance of public health and safety as are current in the atomic energy industry, there would be no aviation or automobile industry, few chemical plants or oil refineries, few sewage systems. Only a governmental industry with unlimited access to the treasury could have met the costs involved in meeting the rigorous standards set.

The standards for occupational exposure to radiation developed by the International Commission on Radiological Protection and the various national protective bodies have stood the test of time. Indeed, they have been adjusted downward in consideration of the more penetrating character of radiations made available in the postwar years, the increased number of persons employed and potentially exposed, and the consequently greater genetic hazards. There is no concrete evidence that any person who has kept his exposure below the radiation protection guides has been injured, although many who have appreciably exceeded these limits have been injured or have died of radiation-induced maladies.

The nonspecific life shortening, the excess leukemia, and the

Editors' note. This paper was received by the editors on April 3, 1970.

excess skin cancer that occurred in a number of the earlier American radiologists have largely disappeared among the more careful practitioners of today and did not appear in excess among their English counterparts after 1921.

However, Tamplin believes on theoretical grounds, arguing largely from application of calculations based on the linear hypothesis, that our protection standards for the general public, one-tenth that for occupational workers, are ten times too high. The linear hypothesis is simply a working one and is used as an aid in the calculation of maximum permissible concentrations for radiation protection. There are many observed exceptions to it in animals and man. It is more probable that one is dealing with a curve with a very gradual initial slope, then a steeper rise (the initiation of this rise being what R. D. Evans has called the practical threshold), and then a tendency to flatten at the top. Also there are other factors that must be considered, as well as radiation, in the induction of tumors. Thus, among some 3,000 white uranium miners in the United States, over 60 developed lung cancer, but nearly all were also heavy cigarette smokers. In over 800 Indian miners, largely Navahos who did not smoke, there has been only one case of lung cancer. In the experimental area at low-level doses, results are often equivocal; in some experiments the animals receiving a small amount of radiation do better than the control animals. Tamplin also emphasizes the potential risk to populations from nuclear plants and, unwarrantedly, lumps together a fuel-processing plant and nuclear power plants. He also implies but does not overtly state that comprehensive studies taking "into account both physical and biological concentrating mechanisms" have not been made. This is in error. Perhaps not every possible nuclide has been followed, but enough were to know the broad patterns and to check carefully those known to have special biomedical risks.

While Tamplin's hypothetical river might not be unreasonable, I doubt that any population could be found that existed totally on a diet of aquatic origin derived from that river. The practical experience after twenty-five years of operation of the Hanford plant, with several reactors and a fuel-processing plant, is that fish caught downstream are eaten by man and have not been dangerous. Tamplin's Table 5.2 giving best estimates of doubling dose of radiation for human cancers and the increase in incidence rate per rad of exposure is not one generally accepted by those interested in carcinogenesis. I know of no data that would permit exact measurement of any increase in incidence rate of cancer per rad.

Merril Eisenbud, who has experience in protection of workers who handle radioactive ores, nuclear fuel, nuclear power plants, and other sources of radiation, tends in his study (Chapter 4) to be perhaps too optimistic. He has presented well the types of reactors and the role of regulatory and standard-setting bodies. As Eisenbud says, there are regions of the earth where the inhabitants are exposed to elevated levels of radiation because of radioactive minerals in the soil, apparently without harm. Studies are still in progress. However, large numbers of these individuals have not been exposed to 100 times the average background level, though small areas with readings at this level can be found. A figure of 10 might be more fair.

Eisenbud properly points out that the AEC requires the licensee of a reactor to carry on monitoring programs in the surrounding area and to take appropriate action if excessive amounts of radioactivity are found. Actually the levels found are usually small fractions of the permissible concentrations. Since these very low levels are attainable at present, there is strong temptation to set the permissible levels comparably low, regardless of the absence of demonstrable deleterious biologic effects at presently set levels and in spite of spiraling costs. Such low levels should not be required simply because they can be achieved. Maximum permissible concentrations should be changed only when the changes can be based on firm evidence.

Eisenbud estimates that the amount of krypton-85 and tritium (H-3) might double by around 2000 A.D. but suggests that the effective doses even then would be in the microrem range, far below the dose for any measurable biologic effect.

The dual role of the AEC in stimulating nuclear power development and in controlling possible effects has led to some apprehension by the public that the agency may not be an adequate watchdog, as Eisenbud says. However, the record of the past quarter century has been good, and AEC's efforts to bring other federal and state agencies into control activities are praiseworthy.

If our power needs and population grow, we must choose between risks and gains in a number of areas, of which this is only one.

7

**Investigation of the
Effects of X-Ray Exposure
of Human Female Fetuses
as Measured by Later
Reproductive Performance:
Interim Summary**

Mary B. Meyer,
Earl L. Diamond,
and Timothy Merz

In his chapter on "Nuclear Reactors and the Public Health and Safety," Arthur Tamplin has referred to our "Investigation of the effects of prenatal x-ray exposure of human oogonia and oocytes as measured by later reproductive performance."[1] As a result of Tamplin's references, the editors invited us to present some comments on our work and some additional data. The following report is not offered here as evidence for or against a particular point of view about the dangers of low doses of radiation to human populations. Our study was never intended as an attack on the use of nuclear fuel for power generation but was designed to investigate a possibly sensitive indicator of damage that might be caused by low doses of x rays in the range currently in use. It should be pointed out that, although it is often said that "we know more about radiation than any other omnipresent form of pollution," there are a few surprises left for us in the bag.

X–ray early in pregnancy is now avoided because of the clearly recognized dangers to the developing fetus,[2] but less attention has been given to the sensitivities of the female fetal gonads and germ cells. The fact that there are no mitotic divisions after seven months of fetal life means that the entire stock of germ

---

Editors' note. The research upon which this publication is based was performed pursuant to contract #CPE-R-69-13 with the U.S. Public Health Service, Department of Health, Education, and Welfare. The manuscript was received by the editors on May 18, 1970.

The references for Chapter 7 are on pages 91–93.

cells is fixed before birth and can only decline thereafter.[3] There are abundant observations from various studies on the susceptibility to radiation damage of both oogonia and oocytes in animals. These observations may be summarized as follows:

1.
It is known that germ cells in different species follow the same pattern of development and that radiation response of these cell stages is a general phenomenon independent of the species of animal. This provides a reliable basis for interspecific comparison of the effects of radiation on reproductive performance, chromosome damage, and mutation frequency.[4]

2.
Two known periods of vulnerability of female germ cells coincide with (a) the final oogonial division, and (b) the diplotene stage of meiosis I and the time of entry into dictyate, if this occurs. Intermediate stages of meiotic prophase appear to be considerably more resistant to the induction of observable damage.[5] In humans, oogonial divisions are observed up to the seventh month of gestation. Cells at diplotene are present from the fourth month of fetal life on, and all oocytes reach this stage around the time of birth.[3] Therefore, the times of increased vulnerability would be expected to occur at 22–30 weeks of gestation and again at 40 or more weeks after conception. This is not a precise definition, since there is considerable overlapping of stages. We cannot say that other times are safe.

3.
The problems of characterizing primary effects by later observations are many. Complicating factors include the induction of superovulation by low or moderate doses of radiation. Superovulation is characterized by an increased shedding of ova early in reproductive life. In rats and mice this event results in a few litters of normal size but is followed by sterility.[6] Other difficulties in assessment are caused by an apparently compensatory leveling off of the natural rate of decrease in the population of

oocytes after birth; the elimination of cells carrying gross chromosomal aberrations or mutations;[5] and the observation that cells may be more sensitive to killing at one stage and to induction of chromosome breakage at another.[7]

4.
Dosage and dose-response relationships are obviously important but may be overshadowed by other factors such as species of animal or the time of exposure. For example, the dose for 50% lethality (LD_{50}) for primary oocytes is 8.4 rads for mice, 100 rads for rats, and 9,000 rads for monkeys.[8] In 21-day-old mice treated with 20 rads, the reproductive capacity is largely destroyed,[9] whereas other species at other times of exposure are not discernibly affected by doses many times larger.

Clearly then, there is sufficient evidence to lead one to ask the following questions: When a pregnant human female is exposed to x-ray pelvimetry, what effect might this cause in her grandchildren? If there is an effect, how may this be observed? The following report provides some evidence in answer to both questions.

The Pelvimetry F_2 Study

The possibility of investigating this problem arose from the existence of data from another study,[10] in which a large number of babies exposed in utero were matched with unexposed babies by race, sex, parity, hospital, and date of birth, using birth records. These children were later located. Many were found in Baltimore schools, and at the start of the pelvimetry F_2 study the oldest ones were seventeen years of age. Our F_1 population comprises 1,500 black females in Baltimore exposed in their own fetal lives and 1,500 matched controls, all born in 1947–1952. Our F_2 babies are the offspring of these 3,000 women or are, in other words, the grandchildren of the women x-rayed in pregnancy. The F_2 births are ascertained from Baltimore birth certificates.

7 The Effects of X-Ray Exposure of Human Female Fetuses

A full report of the study design and analysis of data including births through 1967 has been published.[1] At that time we reported: Exposed mothers had higher birth rates than controls and similar rates of neonatal deaths, fetal deaths, and ascertained abortions. Most comparisons of exposed vs. control F_2 births failed to show a significant increase in problems encountered by exposed mothers in pregnancy and labor, or in the condition and course of their babies. On the other hand, when F_2 births were analyzed by the time during fetal life when the F_1 exposure occurred, the following differences were observed: (1) F_1 mothers exposed before 30 weeks of fetal life gave birth to significantly increased numbers of male babies and to fewer female babies compared with F_1 mothers exposed later in fetal life, or compared with controls. These babies had longer gestations and lower Apgar scores. Placenta weights were lower, and more mothers in this group had precipitate labors. (2) Babies born to mothers exposed at 37 or more weeks of gestation had significantly more neonatal jaundice.

Data summarized in this chapter include additional pregnancies that occurred or were ascertained in 1968 (see Table 7.1). The most striking feature here is that significantly more pregnancies have occurred in the exposed individuals than in the controls. Table 7.1 indicates the components of this difference. Matching or control of slight differences in economic status between the two groups does not alter the finding. Exposed F_1 individuals have had more live births, more fetal deaths, and more abortions. By parity, more have had pregnancies, and those who have had births have had larger families. Neonatal and later death rates and rates of twinning are similar in F_2 exposed and control babies.

Because the stage of development at which exposure occurs bears a crucial relationship to radiation sensitivity, we have analyzed F_2 births by the time of F_1 exposure (see Table 7.2). The increase in birth rates is most marked in those exposed at up to 34

Table 7.1
F_2 Births to Matched Exposed and Control F_1 Mothers
(All births ascertained through 1968)

	Exposed	Control	χ^2	p^a
Number in F_1 population	1458	1458		
F_2 live births	784	673	8.46	.004
F_2 fetal deaths[b]	20	8	5.14	.023
F_2 abortions[c]	30	19	2.47	n.s.[d]
Total pregnancies	834	700	11.70	.001
Twins	14	12		
Deaths—neonatal and later	29	32		
Parity				
First births	529	477	2.68	n.s.
Later births	255	196	7.72	.005

[a] p = probability for chance occurrence.
[b] The Baltimore City Health Department requires a certificate of fetal death for any fetus dead at birth if the period of gestation is 20 or more weeks, or if the fetus weighs 400 or more grams.
[c] Abortions are poorly ascertained by the study methods.
[d] n.s. = not significant.

weeks of gestation and is also elevated in mothers exposed late in gestation (40 or more weeks). Birth rates for mothers exposed at 35–39 weeks, however, do not differ from control rates.

By sex, we find a significant increase in male births and a decrease in female births to F_1 individuals exposed before 30 weeks of gestation. At 30–34 weeks and at 40 or more weeks the female rates are elevated, and the male rates at 40 or more weeks are relatively depressed. At 35–39 weeks, equal numbers of males and females are born.

Discussion

Analysis of data including the 375 additional pregnancies that were ascertained for 1968 corroborates our earlier findings. The data suggest that very low doses of x ray to the female fetus at certain times of gestation have an effect on her offspring. At the early ages observed so far (15 to 21 years old in 1968), exposed

7 The Effects of X-Ray Exposure of Human Female Fetuses

Table 7.2
Pelvimetry F_2 Study—F_2 Births by Weeks of F_1 Gestation at Time of X Ray
(Live births through 1968)

	Weeks of gestation at time of F_1 x ray									All exposed[a]		Controls	
	<30		30–34		35–39		40+						
	No.	Rate[b]	No.	Rate[b]	No.	Rate[b]	No.	Rate[b]		No.	Rate[b]	No.	Rate[b]
F_1 population	111		231		635		294			1458		1458	
Live births	68	61	136	59	303	48	159	54		784	54	673	46
Males (live births)	45	40	62	27	153	24	71	24		390	27	323	22
Females (live births)	23	21	74	32	150	24	88	30		393	27	349	24

[a] Includes individuals with gestation times at time of exposure unknown.
[b] Rates are per 100 in F_1 population.

individuals have had more pregnancies, and the sex ratios of their offspring are altered, with the direction of change depending upon the time of exposure. Increased numbers of fetal deaths and abortions are also observed in exposed F_1 individuals as compared with controls. We do not know yet whether these women will have decreased fertility or become sterile as they grow older.

The effect of ionizing radiation on oocytes is complex, depending upon dose, dose rate, and the stage of development of the cell and of the follicle. Superovulation in animals, induced by relatively low doses of radiation, consists of more eggs being ovulated with the result that litters are larger than normal at conception and equal to control-litter sizes at the time of birth following prenatal death of a number of offspring. This process uses up the available eggs at an increased rate and results in sterility after a few litters. In humans, radiation is known to stimulate ovulation. Some evidence indicates that radiation causes an acceleration of meiotic processes so that, although the number of eggs at early stages is sharply reduced, the number at later stages may temporarily equal or exceed that seen in unexposed controls.[11] A mechanism that might explain the increased number of pregnancies in these young girls is that irradiation of the fetal ovaries and oocytes reduces the proportion of anovulatory cycles usually observed for a year or two after puberty, thus increasing the probability of conception. At ages fifteen to twenty, ovulation is thought to be well established and to occur in 90% of cycles.[12] Consistent with this hypothesis is the observation that the ratio of exposed to control F_2 births is greatest at the youngest maternal ages and approaches unity at ages of nineteen and over (see Table 7.3).

Irradiation of female germ cells is generally expected to result in a decrease in male births because a maternal X chromosome carrying recessive lethal mutations could express itself in a male child (XY), whereas daughters (XX) would be protected by the

Table 7.3
F₂ Births by Exposure and Mother's Age

Mother's age[a]	Number of births		Ratio (exposed:control)
	Exposed	Control	
$\leq$ 14 years	56	41	1.36
15–18 years	591	503	1.17
19–21 years	137	129	1.06
Total	784	673	1.16

[a] F_1 born in 1947–1952. Births through 1968. Data for older ages are therefore incomplete. Ratios would be the same for numbers of babies or for rates, since there are equal numbers of exposed and control F_1 females, matched by birthdate.

normal X chromosome from the father. We observe that F_1 individuals who were exposed at 30–34 weeks or at 40 or more weeks of gestation have had more female than male babies. Although this may reflect damage to the maternal X chromosomes, with a resulting loss of male offspring, we should note that both male and female birth rates are higher than the control rates. It is more difficult to explain the observation that F_1 individuals exposed before 30 weeks of gestation have given birth to almost twice as many males as females, representing a large increase in the male rates and an absolute decrease in female rates. Most of the proposed hypotheses involve gross alterations in the chromosomal material or nondisjunction, and these possibilities can neither be affirmed nor eliminated by the data as yet. It is well known that Turner females with a missing X chromosome (XO) have a poor survival rate as compared with Klinefelter males (XXY). If we assume nondisjunction of the maternal X chromosome and assume that survival of the resulting genotypes will be in proportion to their relative observed frequencies among newborns, we would expect from 100 conceptions 25 XXY males, 12 XXX females, 5 XO females, and no YO males; in other words, 60% male progeny.[13]

To test the hypothesis that these types of individuals were appearing among F_2 children born to mothers exposed in the first 29 weeks of gestation, we attempted to get buccal smears for sex chromatin analysis from offspring of mothers exposed in this period and of their matched controls. Satisfactory specimens were obtained from 28 exposed (43%) and 32 matched control (55%) F_2 babies. All of these had a sex chromatin constitution consistent with their phenotypic sex. The results of this sample suggest that nondisjunction of the maternal sex chromosomes is not a prominent cause, if any, of the observed sex difference in these babies. The possibility of more subtle chromosomal alterations and mutations has not been tested. For example, a number of phenotypically male animals have been observed to have an XX genotype. This has led to the interesting hypothesis that Y material is present in the X chromosome in a repressed state and that a chromosomal disruption might lead to its derepression.[14] This could result in an apparent male with only X chromosomes. The failure to observe sex chromatin in the sample of male children who were tested indicates that they are not simply phenotypic males with two normal X chromosomes. This hypothesis also needs further testing. In addition, as the whole body of the fetus is exposed, we cannot rule out the possibility of hormonal changes that might cause some of the effects we are observing.

To relate these observations more directly to Tamplin's interpretations of our work, we may summarize as follows. Although we have suggested chromosomal damage as a possible mechanism, we need more evidence before we can accept or reject this hypothesis. Except for those who were exposed at 35–39 weeks of fetal life, exposed F_1 females have increased fertility over controls. In the group exposed before 30 weeks of gestation, the excess F_2 births are all males, and there is an absolute reduction in the number of females of a similar magnitude. Therefore, if one postulates the loss of half the female conceptuses, one has also to explain the increase in males. As the sex ratio at conception is unknown and the loss of conceptus is high, perhaps 30% or more, we feel that any theories about what is causing these differences are largely speculative. We think the outstanding feature of our findings is the fact that relatively low doses of radiation to the F_1 female fetus do have a detectable effect on the F_2 generation, whatever the mechanism. This seems to us to be the critical point in relation to setting standards for exposure allowances for the general population.

We are now ascertaining 1969 births to our study population for further analysis and plan follow-up studies of their reproductive success as they reach their twenties and of possible chromosomal damage in their children.

References

1
M. B. Meyer, T. Merz, and E. L. Diamond, "Investigation of the effects of prenatal x-ray exposure of human oogonia and oocytes as measured by later reproductive performance," *Am. J. Epidemiol., 89,* 619–635 (1969).

2
A. S. Dekaban, "Abnormalities in children exposed to x-radiation during various stages of gestation: Tentative time table of radiation injury to the human fetus, Part I," *J. Nuclear Med., 9,* 471–477 (1968).

3
T. G. Baker, "A quantitative and cytological study of germ cells in human ovaries," *Proc. Roy. Soc. B, 158,* 417–433 (1963).

4
E. F. Oakberg, "The influence of germ cell stage on reproductive and genetic effects of radiation in mammals," *Dis. of Nervous System,* Monograph Supplement, *24,* No. 4 (April 1963).

5
A. M. Mandl and H. M. Beaumont, "The differential radiosensitivity of oogonia and oocytes at different developmental stages," in *Effects of Ionizing Radiation on the Reproductive*

System, W. D. Carlson and F. X. Gassner, eds. (New York: Pergamon Press, Macmillan, 1964), pp. 311–321.

6

L. B. Russell and W. L. Russell, "The sensitivity of different stages in oogenesis to the radiation induction of dominant lethals and other changes in the mouse," *Proc. Roy. Soc. B, 158,* 187–195 (1963).

7

L. B. Russell and C. L. Saylors, "The relative sensitivity of various germ cell stages of the mouse to radiation-induced nondisjunction, chromosome losses and deficiencies," in *Repair from Genetic Radiation Damage and Differential Radiosensitivity in Germ Cells,* F. H. Sobels, ed. (New York: Macmillan, 1963), pp. 313–340.

8

S. Zuckerman, "The sensitivity of the gonads to radiation," *Clin. Radiol., 16,* 1–15 (1965).

9

H. Peters and E. Levy, "Effect of irradiation in infancy on the fertility of female mice," *Rad. Res., 18,* 421–428 (1963).

10

A. Lilienfeld, "Epidemiologic studies of the leukemogenic effects of radiation," *Yale J. Biol. and Med., 39,* 143–164 (1966).

11

P. Vorisek and J. Vondracek, "Der Mechanismus der postnatalen Veränderungen des Follikelapparates bei intrauterin mit kleinen Dosen bestrahlten Ovarien," *Strahlentherapie, 135,* 602–609 (1968).

12

P. Eckstein, "Ovarian physiology in the non-pregnant female," in *The Ovary,* Vol. I., S. Zuckerman, A. M. Mandl, and P. Eckstein, eds. (New York and London: Academic Press, 1962), 311–359.

13

V. Mittwoch, "Sex chromatin," *J. Med. Genetics, 1,* 50–78 (1964).

14
J. D. Biggers, "Aspects of intersexuality in domestic mammals," *VIe Cong. Intern. Reprod. Anim. Insem. Artif., 11,* 841–870 (1968).

8

Radiation Dose Limits   Study Panel on
Nuclear Power Plants,
Maryland Academy of Sciences

Introduction

The establishment by the Maryland Academy of Sciences of its Study Panel on Nuclear Power Plants was occasioned by a great deal of public controversy in Maryland concerning construction of a large nuclear power plant at Calvert Cliffs on Chesapeake Bay by the Baltimore Gas and Electric Company.

From May 1969 until January 1970, the Study Panel met frequently, interviewed scientists, engineers, conservationists, economists, and government officials, studied all available information, and finally published a 48-page report of which the following section is a part.

In its report, the Study Panel stated that it considered the Calvert Cliffs plant to be "an acceptable risk" when looked upon as "an experimental tool." It then went on to make the following specific recommendations:

1.
That any potential user of the water of the Chesapeake Bay must perform ecological and oceanographical studies as directed by the proposed authority and that these studies be made public.

2.
That the budgets of the Maryland Departments of Water Resources and Health and Mental Hygiene be increased "immediately and substantially in order to provide for adequate chemical, thermal, microbiological and radiation monitoring of

Editors' note. This chapter was originally published in *Nuclear Power Plants and Our Environment*, a Report to the Maryland Academy of Sciences by the Study Panel on Nuclear Power Plants (Baltimore, Maryland: January, 1970), pp. 36–44. It is reproduced with permission, with minor editorial changes. References for Chapter 8 are on pages 104–106.

all substantial effluents discharged into the Maryland environment including those from nuclear power plants."

3.

That Maryland continue its policy against permitting reprocessing or disposal of spent nuclear fuels within the state.

4.

That the Atomic Energy Commission be asked to make a more stringent allocation to the nuclear power industry of radiation dose limits for the general public.

Owen Phillips, professor of earth and planetary sciences at The Johns Hopkins University, chaired the panel during the eight-month period. Other panel members were Howard Laster, chairman of the Department of Physics and Astronomy, University of Maryland; Howard H. Seliger, Department of Biology, The Johns Hopkins University; and J. W. Foerster, Department of Biology, Goucher College, Ex officio members of the panel were Robert P. Rich, Academy president; Robert L. deHaan, Academy vice-president; and Nigel O'C. Wolff, Academy director.

The Report

At the present time the legally defined limits of exposure to ionizing radiation are incorporated in a statement of radiation protection guidance published in the Federal Register by the U.S. Federal Radiation Council (FRC).[1] In order that we understand the radiation protection guidelines and their intent and application to radiation workers, to individuals in the population, and to averages over large numbers of persons in the entire population, it would be best to discuss the reasons for permitting anyone to be exposed to radiation in excess of natural background.

Natural Background Exposure

The irreducible exposure of populations in the United States to natural background radiation amounts to between 0.1–0.2 rem per year, depending on geographical location.[2] Assuming that the mean age for childbearing is thirty years, this gives a background genetic dose of 3–6 rem. It is this irreducible minimum against which all other exposure must be weighed.

Medical Exposure
It has been generally accepted that medical exposure to x rays for diagnostic purposes and use of ionizing radiation for the treatment of disease have benefits to the individual that exceed the risks of radiation damage involved. For this reason and to avoid any arbitrary limitation on the medical profession in its use of ionizing radiation, neither the national and international organizations on radiation protection nor the FRC has included medical exposure in their recommendations of dose limits, although concern is expressed that medical radiological procedures be limited to the minimum consistent with medical benefit. It is beyond the scope of this report to assess whether in actual practice individual patients or doctors have been unnecessarily overexposed as a consequence of poor diagnostic procedures and a general lack of awareness of the hazards of radiation.[3]

Earlier estimates of the genetically significant dose stemming from medical diagnostic exposure were as high as 0.14 rem per year.[4] More recent estimates place this exposure at 0.055 rem per year. It is the considered judgment of many acknowledged experts in diagnostic radiology that this genetic dose should and can easily be reduced by a factor of 10 to 0.005 rem yer year or 0.15 rem per 30 years.

Additional Exposure
The inclusion of the medical exposure genetic dose and the implication that it can be reduced significantly is intended only as a marker of a "voluntary" exposure so that we may have some reference point for assessing the "involuntary" exposure to which

the general population may be exposed as the result of peaceful uses of atomic energy.

The question therefore arises: What additional genetic dose should the general population be exposed to, based on the premise that excess radiation produces harmful effects, both to present and future generations? The difficulty up to the present time is that the effects are so complex and contain so many unknown factors that the assessment of quantitative damage is subject to large uncertainties.

It is the general feeling among human population geneticists that only the order of risk associated with radiation exposure can be estimated. For example, a fifth-order risk implies that the probability of an event (death or injury) to any individual in the population is in the range 1×10^{-5} to 10×10^{-5}. Thus a fifth order risk per year for thyroid carcinoma in the United States (population 200 million) would result in the appearance of 2,000–20,000 cases per year. The International Commission on Radiological Protection Publication 8 presents data for both somatic and genetic effects of radiation exposure.[5] For the purposes of this proposal, only the conservative assumptions of linearity of damage with dose and independence of damage with dose rate will be used.

It is important to qualify the term "genetic death" in establishing an order of risk: "A genetic death is the extinction of a gene lineage through the premature death or reduced fertility of some individual carrying the gene." [6] In order that there be no uncertainty about the origin of figures used for risk calculations, Tables 13 and 15 of ICRP Publication 8 are reproduced here (as Figures 8.1 and 8.2).

The present legal genetic dose of 5 rem corresponds to a dose rate of 0.17 rem/yr for 30 years. From Figure 8.1, considering that the parental generation has received 5 rem, the predicted number of genetic deaths per million births will be 1,055. Based on yearly births of 4.6 million, the annual genetic deaths are

Table 13. *Estimated detriment to first generation offspring in a population where the whole parential generation has received 1 rad or 30 rads*

Type of detriment	Number expected without parental man-made irradiation	Number estimated to have arisen in preceding generations without parental irradiation	Estimated* additional numbers resulting from	
			1 rad	30 rads
A Autosomal dominant gene traits (births)	8000[a]	320[a]	16[a]	480
B Sex linked traits (births)	250[b]	<83[b]	<4[b] (Mother exposed)	<120
C Chromosomal aberrations (births)	7000[c]	7000[c]	Unknown[c]	Unknown[c]
D Abortions associated with chromosomal aberrations (recognized pregnancies)	35,000[c] (i.e. 20% of all abortions)	35,000[c]	Unknown[c]	Unknown[c]
E "Genetic deaths" (zygotes)	235,000[d]	5875[e]	211[f]	6330

* The number of effects anticipated per million births are shown; the total number of pregnancies in the population which would last long enough to be recognized would be about 1,175,000, there being about 175,000 abortions of which about 20 per cent are assumed to be associated with chromosomal aberrations. To facilitate alternative calculations figures have not been rounded and the number of significant digits is not indicative of precision.

[a] 0.8 per cent have dominant gene traits, i.e. 8000 per million. The proportion of sporadic cases is 4 per cent, i.e. 320. Both parents receive irradiation so the number expected from 1 rad would be 1/20th if the doubling dose is 20 rads, i.e. 16.

[b] From footnote (b) to Table 12 the birth frequencies of those traits is 5×10^{-4} in male births, i.e. 250 in 1,000,000 births of both sexes. If $\frac{1}{3}$ of these were mutants, we should expect 83 mutant subjects. Increase in the next generation by 1/20th would give an expected 4 cases.

[c] No predictions are made (see Section 10.1.5) but current frequencies are shown to draw attention to their magnitude.

[d] If the average individual carries 8 small dominant mutations each conferring independent risks of 2.5 per cent, then the total risk is $0.025 \times 8 = 0.2$. Therefore, in 1,175,000 zygotes we should expect 235,000 eliminations per generation. By postulating equilibrium gene frequencies, a figure is arrived at whereby the same number of mutations which arise in each generation are eliminated; otherwise the total gene frequency would either rise or fall.

[e] If of these 235,000 newly arisen mutations 2.5 per cent are eliminated in the first generation, then the number eliminated would be $0.025 \times 235,000 = 5875$.

[f] If 1,175,000 zygotes received 2,350,000 gametes from their parents and if these parents received an average 1 rad of radiation, then the mutations received by their offspring would be $2,350,000 \times 0.0036 = 8460$ mutations. Of these $0.025 \times 8460 = 211$ would be expected to be eliminated in the first generation as "genetic deaths".

8.1
Photographic reproduction of Table 13 in *The Evaluation of Risks from Radiation,* ICRP Publication 8, a report prepared for Committee I of the International Commission on Radiological Protection (New York, London: Pergamon Press, 1966), p. 46. The table is reproduced with the permission of the publisher.

calculated to be 4,850, and the annual risk (normalizing to 200 million population) is therefore 2.4×10^{-5}. From Figure 8.2 the annual risk of all induced cancers under continuous exposure of 0.17 rem per year is 1.2×10^{-5}.

K. Z. Morgan has provided independent estimates of annual

Table 15. *Estimated risk of cancer from exposure to 1 rad if a linear dose/response relationship below 100 rads is assumed*

Type of detriment		Estimated cases per 10^6 exposed persons*	Order of risk to the individual
Fatal neoplasms	⎰ Leukaemia†	20	5th
	⎱ Others	20	5th
Thyroid carcinoma‡		10–20	5th

* Effects would be experienced over 10–20 years.
† Risk may be enhanced by factor of 2–10 if the foetus is exposed.
‡ Estimate refers to exposure in childhood; unlike other classes of cancer for which estimates are given incidence is not equivalent to mortality.

8.2
Photographic reproduction of Table 15 in *The Evaluation of Risks from Radiation,* ICRP Publication 8, a report prepared for Committee I of the International Commission on Radiological Protection (New York, London: Pergamon Press, 1966), p. 56. The table is reproduced with the permission of the publisher.

mortality due to genetic death and cancer.[3] Using Morgan's figures, modified to an annual dose rate of 0.17 rem, the following is calculated:

Annual risk of genetic death: 3×10^{-5} to 4×10^{-4}/yr
Annual risk of cancer: 2×10^{-5} to 5×10^{-5}/yr

T. C. Carter[7] has advocated a different criterion from genetic death. He has considered the effects of radiation-induced genetic damage which produces the socially handicapped. Normalizing his totals (from his Table V) to a genetic dose of 5 rem, there would be a predicted increase of 1,900 persons born with various genetic handicaps per million persons exposed. Based on a yearly birth of 4.6 million, the annual risk of genetic handicap is 4.4×10^{-5}/yr.

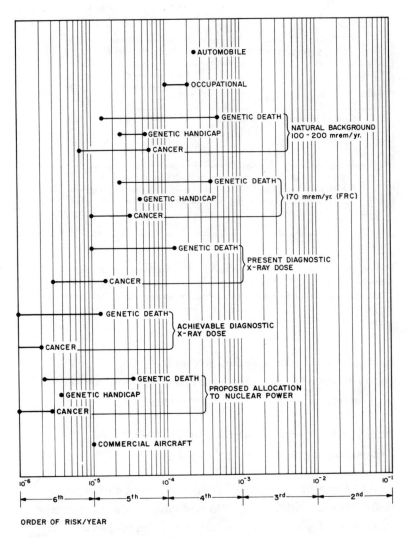

8.3
Comparison of the risks from radiation with other, more familiar risks.

In order to establish a frame of reference within which to weigh these risks, a graph illustrating various risks with which we may be more familiar would place risks from radiation in some perspective. Figure 8.3 indicates the ranges calculated from the

ICRP data,[5] Morgan's data,[3] and Carter's data.[7] The interesting result is that the most conservative estimates of genetic death risks from natural background, the present diagnostic x-ray dose, and the present FRC limit for genetic dose from nuclear power are all fourth-order risks, along with the automobile.

There are quite valid objections to the comparison of genetic death risks with other types of risks. In addition, the natural genetic deaths as shown in Figure 8.1 are already very high. One could make a more cogent argument by using the data for genetic handicap or induction of cancer (somatic risk), since the comparisons with the other types of risks are more related. In both of these cases the risk from natural background is a fifth-order risk.

Expert testimony has been introduced[3] showing that it is possible, with the technology now available, to reduce somatic risks from diagnostic medical x-ray exposure from a fifth-order risk to a sixth-order risk or below. If we use this "voluntary" medical exposure as a guide, there seems to be no reason why the "involuntary" population somatic exposure risk from radioactive waste from the nuclear power industry should not also be reduced to a sixth-order risk or below.

It is the nature of the radioactive material produced by nuclear reactors that, once the nuclides are released to the environment, many remain for generations even if all further production is stopped. It is therefore critical that only the most conservative estimates of damage be applied.

With the possible exception of critical organ radionuclides, the reduction of permissible exposure levels to the general population to limit somatic risk from radioactive waste to a sixth-order risk or below would also reduce the genetic dose by approximately the same factor. For this reason, the term "genetic dose" will be used even though arguments are based on the somatic effects of this whole body dose.

In ICRP Publication 9, the recommendation is made for a provisional limit of 5 rem for the genetic dose to the general population from all sources additional to natural background radiation and medical exposures.[8] The following statement illustrates the trepidation felt about even partially approaching this value: "It should be emphasized that the limit may not in fact represent a proper balance between possible harm and probable benefit, because of the uncertainty in assessing the risks and the benefits that would justify the exposure." This maximum population genetic dose is carried in the FRC's radiation protection guidance[1] — that is, for exposed populations a suitable sample should not exceed 0.17 rem per year. This figure is the mean dose rate over a 30-year period corresponding to the present legal maximum genetic dose of 5 rem per 30 years.

There are several maximum levels of exposure the Atomic Energy Commission has set up which are designated in the Code of Federal Regulations.[9] These are 5 rem per year for radiation workers over the age of eighteen, 0.5 rem per year for individuals in the general population, and 0.17 rem per year for suitable samples of exposed populations. Since these regulations are the basis for regulation of the nuclear power industry, it would appear that the industry could legally release sufficient radioactivity to the environment to produce a mean dose rate to the population as high as 0.17 rem/yr.

Since the Atomic Energy Commission, in its public appraisals of the radioactivity released by nuclear power plants, has indicated that present releases of radioactive waste to the environment are only about 1% of the present legal maximums, it is felt that much more realistic constraints can be placed on the nuclear power industry without interfering with the growth of this industry.

It is therefore proposed that the nuclear power industry be allocated a portion of the total dose suggested by ICRP and FRC, not to exceed one-tenth of that stemming from the average natural background. This would permit a total body or gonadal dose not to exceed 0.17 rem per year, corresponding to a genetic dose of 0.5 rem per 30 years.*

At the present time the maximum permissible genetic dose allocated to all types of radioactive waste disposal in the United Kingdom is 1 rem per 30 years. The committee that considered this matter expected that the figure could be kept as low as 0.1 rem per 30 years.[10] The medical exposure genetic dose in the United Kingdom is at present 0.42 rem per 30 years, and a study by G. M. Ardran[11] showed that if all installations used the x-ray diagnostic procedures at present in use in 25% of the installations, the population gonad dose would probably be reduced by a factor of 7 to less than 0.1 rem per 30 years. The American medical profession will very likely reduce the diagnostic x-ray genetic dose to less than 0.2 rem per 30 years. If the present strict regulation of nuclear power waste disposal is maintained, the genetic dose from this radioactive waste will remain well below 0.5 rem per 30 years.

An important criticism of the original suggestion was that it appeared as though the 5 rem per 30-year limit suggested by ICRP and adopted by the FRC was being challenged. Such is not the intent. This figure is a consensus of expert knowledge at the present time; it may be changed as new data become available. More precisely, the issue is whether the nuclear power industry should be permitted, if conditions warrant, to utilize this 5-rem limit or whether a more realistic and lower limit of 0.5 rem should be allocated to nuclear power, which is not the only source of radiation dose to the population.

---

* An intial proposal was distributed informally to a group of persons with recognized interests in radiation dose limits. This revised proposal is based on a response to a request for comments and particularly the extremely helpful and informative suggestions of H. J. Dunster of the United Kingdom Atomic Energy Authority and K. Z. Morgan of the Oak Ridge National Laboratory.

In the context of a small number of nuclear installations — weapons production, research, and the production of radioisotopes for research applications — large populations could not conceivably be exposed, and therefore the potential genetic and somatic load to the total population is insignificant. However, the rapid multiplication of these nuclear facilities and the planned proliferation of nuclear reactors, with the consequent problems of fuel handling, reprocessing, waste disposal, and accidental releases, force upon us the more realistic picture of large numbers of plants and exposures to larger segments of the population.

References

1
Federal Radiation Council, "Radiation protection guidance for Federal Agencies, Memorandum for the President," notices in the *Federal Register, 25,* 4402–4403 (May 18, 1960), and *Federal Register, 26,* 9057–9058 (September 26, 1961).

2
K. Z. Morgan and J. E. Turner, eds., *Principles of Radiation Protection, a Textbook of Health Physics* (New York, London, Sydney: John Wiley & Sons, 1967).

3
Karl Z. Morgan, testimony and prepared statement, "Reduction of unnecessary medical exposure," in *Radiation Control for Health and Safety Act of 1967,* Hearings before the U.S. Senate Committee on Commerce, 90th Congress, 1st session, on S.2067, August 28, 29, and 30, 1967, Serial No. 90–49 (Washington: U.S. Government Printing Office, 1968), pp. 31–64; and Karl Z. Morgan, "Ionizing radiation: Benefits versus risks," in *Radiation Control for Health and Safety Act of 1967,* Hearings before the U.S. Senate Committee on Commerce, 90th Congress, 2nd session, on S.2067, S.3211, and H.R.10790, Part 2, May 6, 8, 9, 13, and 15, 1968, Serial No. 90–49 (Washington: U.S. Government Printing Office, 1968), pp. 871–874, statement prepared for presentation

at the annual meeting of the Health Physics Society, June 16–20, 1968, in Denver, Colorado.

4

John S. Laughlin and Ira Pullman, "Gonadal dose produced by the medical use of x-rays," a preliminary edition of Section III of *The Genetically Significant Radiation Dose Received by the Population of the United States,* published separately as one of the series *The Biological Effects of Atomic Radiation* (Washington: National Academy of Sciences — National Research Council, March 1957).

5

International Commission on Radiological Protection Publication 8, *The Evaluation of Risks from Radiation,* a report prepared for Committee I of the ICRP (New York, London: Pergamon Press, 1966).

6

Ibid., p. 26.

7

T. C. Carter, "Ionizing radiation and the socially handicapped," *Brit. J. Rad., 30,* 641–647 (1957).

8

International Commission on Radiological Protection Publication 9, *Recommendations of the International Commission on Radiological Protection (adopted September 17, 1965)* (New York, London: Pergamon Press, 1966).

9

Code of Federal Regulations (10 CFR 20), Title 10 — Atomic Energy, Part 20 — Standards for Protection against Radiation (Washington: Office of the Federal Register, National Archives and Records Service, General Services Administration, revised as of January 1, 1969), pp. 134–158.

10

Personal communication, H. J. Dunster (UK Atomic Energy Authority) to H. H. Seliger (The Johns Hopkins University).

11
G. M. Ardran, as reported by S. K. Stephenson in "Report of meeting: Hazards and doses to the whole population from ionising radiations, Thursday, 13 May 1965," *Annals of Occupational Hygiene, 9,* 83–88 (1966).

9

Radiation Doses from Fossil-Fuel and Nuclear Power Plants

James E. Martin,
Ernest D. Harward,
and Donald T. Oakley

Trace quantities of uranium and thorium and their products of radioactive decay are released to the environment in the form of fly ash from large fossil-fuel steam electric stations. This poses the question of the significance of these releases compared with those from nuclear power plants that routinely discharge low-level radioactive waste to the environment. It is the purpose of this chapter to compare the releases using data that were obtained by the Bureau of Radiological Health from field studies at operating coal- and oil-fired steam stations and at nuclear power plants.[1]

This comparison is made not with any suggestion that a radiological hazard exists but only to show how human exposures produced by such discharges vary with different types of plants and how these variations are influenced by design parameters. Radiation exposure dose rates were found to be low from all plants; the highest exposure rates were only 1%–3% of the radiation protection guidelines recommended by the Federal Radiation Council for individual members of the population. Likewise, there is no intention to suggest that other factors, such as mining and processing of the fuel or the generation and release of other types of pollutants, should not be considered. All of these factors, of course, enter into any complete determination of the total environmental burden represented; however, because of the unavailability of a common comparison index, this comparison was limited to radiological considerations only.

Editors' note. This paper was received by the editors on April 1, 1970. The references for Chapter 9 are on pages 123–125.

Background Information

Several authors have reported the results of measurements of naturally occurring uranium, radium, and thorium in the residues of fly ash from the combustion of fossil fuels. The results of these measurements are shown in Table 9.1. 2–5 Each group of workers concluded that the alpha radioactivity in emitted fly ash was not of public health significance.

Eisenbud and Petrow compared the relative biological significance of fossil-fuel radioactivity with emissions from nuclear

Table 9.1
Analyses of Radioactivity in Coal and Oil Fly Ash

Sample	Reference	Concentration (pCi/g dry fly ash)			
		Ra-226	Ra-228	Th-228	Th-232
Appalachian coal ash	(4)	3.8	2.4	2.6	—
Utah coal ash	(4)	1.3	0.8	1.0	—
Wyoming coal ash	(4)	—	1.3	1.6	—
Japan coal ash	(4)	—	1.5	1.6	—
Alabama coal ash	(4)	2.3	2.2	2.3	—
Venezuela petroleum ash	(4)	0.21	0.49	0.67	—
TVA coal plants	(3)	4.25	2.85	2.85	2.85
Coal ash (Australia)	(2)	7.98	—	—	—
Oil fly ash[a] Turkey Point	(5)	0.18	0.17	0.82	0.17
Coal fly ash[b] Hartsville	(5)	2.3	3.1	—	3.1
Coal fly ash[c] Colbert, TVA	(5)	3.1	6.9	1.6	6.9
Coal fly ash[d] Widows Creek, TVA	(5)	1.6	2.7	2.8	2.7

Note. pCi = picocuries = 10^{-12} curies; 1 curie = 3.7×10^{10} disintegrations per second. See the Glossary on page 412 for a more complete explanation of the nuclear units.
[a] Average of 6 samples for Ra-226, Th-228, and Th-232; Ra-228 assumed in equilibrium with Th-232.
[b] Average of 5 samples for Ra-226, Th-228, and Th-232; Ra-228 assumed in equilibrium with Th-232.
[c] Average of 12 samples; Ra-228 assumed in equilibrium with Th-232.
[d] Average of 26 samples for Ra-226; Ra-228 assumed in equilibrium with Th-232.

plants and concluded "that an electrical generating station that derives its thermal energy from such fuels discharges relatively greater quantities of radioactive substances into the atmosphere than many power plants that derive their heat from nuclear energy." [4] This conclusion is based primarily on a calculation that a typical 1,000-megawatt coal-fired plant would release 28 millicuries (mCi) of radium per year, which corresponds, on the basis of maximum permissible concentration published by the International Commission on Radiological Protection, to 10,000 and 10 curies of krypton-85 and iodine-131 (two common radionuclides released from nuclear reactors), respectively. Operating experience available to them at that time (1964) for the Yankee Atomic Power Plant (a pressurized water reactor) indicated that actual quantities of krypton-85 and iodine-131 released annually were only small fractions of these values. This conclusion is modified by more recent data reported here from environmental studies at operating nuclear and fossil-fuel plants and more recent discharge data from operating nuclear plants.

Field Studies
Emissions of radioactivity and environmental radioactivity levels produced by these emissions were measured at both operating fossil-fuel power plants and operating nuclear power plants. Data for oil-fired plants were obtained from a two-year study at the Florida Power and Light Company's Turkey Point site. Environmental radioactivity levels produced by emitted oil fly ash were so low that detection was extremely difficult. Since it was established that coal fly ash contained about 10 times more radium-226 than oil fly ash (see Table 9.1), and that the higher ash content of coal results in greater quantities of fly ash being discharged, a greater emphasis was justified for coal plants than for the oil plants. For example, a 1,000-megawatt oil plant with 0.5% ash content and 84% fly ash collection will release about 0.105×10^{10} grams per year (g/yr) of fly ash containing about

1.1 mCi. A similar sized coal plant with 13% ash content will release to the environment about 7.6×10^{10} g/yr of fly ash containing 970 mCi, or about 885 times more radioactivity per megawatt than an oil plant.

Three different coal-fired stations were studied. The greatest emphasis centered on the Widows Creek Station operated by the Tennessee Valley Authority, because it was large (1,960 MW), had older units with less efficient air-cleaning equipment, and had short stacks that would result in poor atmospheric diffusion of stack discharges and was typical of older, less efficient plants elsewhere.

Emission data from several nuclear power plants were obtained through operating experience and from results of a special study conducted by the Bureau of Radiological Health at the Dresden Nuclear Power Station operated by the Commonwealth Edison Company at Morris, Illinois. This study was conducted to improve surveillance programs by measuring the types and quantities of radionuclides released and the resulting environmental radioactivity levels. These data, although taken for a different purpose, were readily comparable with the field measurements at Widows Creek.

Data and Observations

Fossil-Fuel Plants It was concluded from data obtained at various stages of the field studies that airborne radioactivity was the only population exposure pathway in which radioactivity from both fossil-fuel and nuclear power plants was detectable.[1,5,6,7] Although soil might be expected to show a long-term buildup of fossil-fuel radioactivity, it was not observed. This is attributed to the normal variability of soil, the inherent difficulty in obtaining reproducible analytical data for soil, and the specific activity of fly ash, which is about the same as that of soil precluding positive determination of fly ash radioactivity in the presence of normal soil background radioactivity. Since fly ash is rather insoluble, its biological availability is very low and any accumulation on soil was not expected to be and was not, in fact, detected in either vegetation or surface water above the level present from the natural soil background.

Air sampling at the Turkey Point site for airborne fly ash was unsuccessful primarily because filter samples collected monthly by low-volume samplers accumulated substantial airborne dust, the specific activity of which is near that of oil fly ash.[7] In contrast, air sampling at the Widows Creek site utilized high-volume samplers and special techniques to reduce the effect of the airborne dust background and was more successful. The results of these samples are shown in Table 9.2.

Table 9.2
Airborne Radioactivity Concentrations and Dose Rates at the Widows Creek Plant on 5/13/69 [a]

Station no.	Location		Air concentration (μCi/cm^3 × 10^{-15})			Dose rate (μrem/h) [c]	
	Azimuth	Distance	Ra-226	Th-232	U-238	Bone	Lung
1	34°	1.7 miles	1.3	n.d.[d]	2.72	6.0	1.0
2	45°	1.6 miles	0.64	0.26	1.17	3.0	0.8
3	23°	3.9 miles	0.31	n.d.	2.55	1.6	0.4
4	34°	3.7 miles	0.52	0.24	2.54	2.8	0.8
5	43°	3.6 miles	0.64	0.47	1.24	3.1	1.3
6	51°	3.7 miles	0.64	0.33	4.42	3.1	1.2
7	39°	5.1 miles	0.40	0.08	1.05	1.9	0.4
8[b]	136°	2.6 miles	b	b	b	b	b

Note. μCi = microcuries = 10^{-6} curies; μrem = 10^{-6} rem. See the Glossary on page 412 for a more complete explanation of the nuclear units. The figures in the table have been corrected for the observed background readings of clean filters.
[a] Obtained for a seven-hour period during neutral atmospheric stability.
[b] Station No. 8 was placed out of the plume to measure airborne dust. Air concentrations reflect subtraction of these values and filter background levels.
[c] These data are hourly exposure rates for one set of meteorological conditions and are not directly relatable to average annual exposures without accounting for varying meteorological conditions. Bone doses assumed fly ash was soluble; lung doses assumed it was insoluble.
[d] n.d. Nondetectable above filter and normal dust background.

Seven of the station's eight units were at 95% capacity during the sampling period; the eighth unit was shut down. Units 1 through 6 (140 MW each) collectively emitted fly ash at 20,700 lb/h, and Unit 7 (575 MW) emitted 12,000 lb/h.[8] The emissions from Units 1 through 6 were assumed to account for most of the radioactivity collected by the air samplers since the tall stack of Unit 7 (500 ft) and buoyancy most likely caused its plume to reach ground level considerably downwind of the sampling points. On this basis, the measured concentrations were produced by a release rate of 4,160 pCi/sec of radium-226 and thorium-230; 7,150 pCi/sec of thorium-232, thorium-228, and radium-228; and 7,800 pCi/sec of uranium (20,700 lb/h of fly ash containing radioactivity concentrations shown in Table 9.1; pCi = picocuries = 10^{-12} curies).

The most significant radionuclides for exposure from the coal plant were determined by examining the nuclides in the uranium and thorium series relative to the maximum permissible concentrations (MPCs) for soluble and insoluble nuclides recommended by the International Commission on Radiological Protection (ICRP).[9] If the nuclides in fly ash are soluble, bone is the critical organ; if insoluble, the lung is the critical organ. For bone the most significant nuclides in fly ash are thorium-230, thorium-228, thorium-232, radium-226, and radium-228; for the lung they are thorium-230, thorium-232, uranium-238, thorium-228, and radium-228. Bone and lung dose rates were calculated for these nuclides from the field measurements and are shown in Table 9.2. These calculations are for neutral atmospheric stability and assume that an exposed individual possesses the characteristics of the so-called standard man. Of the air concentration values used, only thorium-232, uranium-238, and radium-226 were actually measured. The others were assumed to be in secular equilibrium in the decay series. Thus radium-226 and thorium-230 exist in equivalent amounts in the uranium series;

9 Radiation Doses from Fossil-Fuel and Nuclear Power Plants   113

thorium-232, thorium-228, and radium-228 are present in equal amounts in the thorium series.

Each dose rate in Table 9.2 was calculated using MPCs recommended by the ICRP. For lung exposure, the MPC for unrestricted areas corresponds to a dose rate of 171 microrem per hour (μrem/h) (1.5 rem/yr) after sufficient time for an equilibrium concentration to occur in the lung tissue. For bone, continuous exposure to the MPC in air for unrestricted areas for 50 years would produce a dose rate equivalent to that from a body burden of 0.01 microgram of radium, or 333 μrem/h (0.056 rem/week). Therefore, the actual bone dose rates are very low at the beginning of exposure to airborne fly ash radioactivity and reach the body burden only after 50 years. For purposes of comparison, however, it is assumed that the concentrations observed at Widows Creek represent the fiftieth year of exposure and, therefore, are a percentage of the maximum permissible dose rate recommended by the ICRP. This assumption is conservative for all exposure times less than 50 years.

Nuclear Plants   Studies at Dresden 1 show that the most significant pathway for public exposure to radioactive nuclides is from noble gases emitted from the stack. Several sets of measurements were made in the downwind plumes on five different occasions and during three different atmospheric stability conditions. Data for neutral conditions (similar to those at Widows Creek on May 13, 1969) are contained in Table 9.3 and are averages of several measurements made over periods of 30 to 60 minutes with a tissue-equivalent ionization chamber and Shonka electrometer.[6]

Since both the Dresden and Widows Creek measurements were made during approximately the same meteorological conditions, and the Dresden stack at 300 feet is reasonably comparable to those of Units 1 through 6 (170 ft–270 ft), the resulting exposure dose rates were assumed to be comparable. This com-

Table 9.3
Measured Dose Rates in Discharge Plumes from Dresden 1

	Location		Stability condition	Release rate	Dose rate[a]
Date	Distance	Azimuth			
1/16/68	1.2 miles	190°	Neutral	14,000 μCi/sec	32 μrem/h
1/17/68	1.1 miles	192°	Neutral	15,000 μCi/sec	40 μrem/h
8/21/68	1.1 miles	39°	Neutral	11,600 μCi/sec	24 μrem/h

[a] These are time-average plume dose rates for periods of 30 minutes to 1 hour; any extrapolation of these data to annual doses must account for varying meteorological conditions.

parison was based on the dose limits for various critical organs recommended by the ICRP, assuming the risk factors for various modes of exposure are the same and are reflected in the ICRP values. Whole body exposure of individuals in the public from noble gases is limited to 0.5 rem per year; bone dose from long-lived bone seekers is limited to that from 0.01 microgram of radium or 2.9 rem per year; and lung doses from insoluble particles are limited to 1.5 rem per year. Hourly exposure rates to produce these annual doses were calculated as a basis for comparing the field data.

The relative significance of the measured dose rates for the nuclear plant and of those calculated from air measurements at the coal plant was determined by calculating their respective fractions of the recommended limit. These fractions were then related to each other to find which type of exposure was the greatest in terms of the recommended ICRP value. From Tables 9.2 and 9.3 the following observations can be made:

1.
The maximum bone dose rate from Widows Creek was 6.0 μrem/h for station 1 or 1.8% of the ICRP value of 333 μrem/h for unrestricted areas.

2.
The highest dose rate for lung exposure was observed at Station 5 to be 1.28 μrem/h or 0.7% of the ICRP value of 171 μrem/h.

3.

The average whole body dose rate above background measured from the plumes at Dresden 1 was 30 μrem/h, which is 53% of the ICRP hourly limit of 57 μrem/h (0.5 rem/yr) for whole body exposure.

4.

On an adjusted basis of equal power-generating capacity, the Dresden 1 boiling water reactor represents an exposure dose-rate that is about 120 times that from Widows Creek for soluble fly ash (bone dose) or about 300 times greater for the insoluble fly ash (lung dose).

(These values are based on limited field data and were compared for fixed meteorological conditions. They would be substantially reduced when extrapolated to annual average exposure rates because of meteorological variations.)

Discharge data for selected nuclear power reactors reported for 1968[10] are presented in Table 9.4 to show how other operating reactors would compare to a coal plant such as Widows Creek.

Table 9.4
Releases of Noble and Activation Gases from Selected Power Reactors, 1968

Facility	Type	Power (MW)	Total discharge (Ci)	Avg. release rate (μCi/sec)	% of Dresden 1
Dresden 1	BWR[a]	200	240,000	7,600	100
Humboldt Bay	BWR	69	897,000	28,200	370
Big Rock Pt.	BWR	70	232,000	7,350	97
Connecticut Yankee	PWR[b]	462	3.70	0.12	0.0017
San Onofre	PWR	430	4.75	0.15	0.0020
Indian Pt. 1	PWR	265	55.20	1.75	0.0230
Yankee	PWR	175	0.66	0.02	0.0003

[a] Boiling water reactor.
[b] Pressurized water reactor.
Source. Joint Committee on Atomic Energy, 91st Congress, 1st session, *Selected Materials on Environmental Effects of Producing Electric Power* (Washington: U.S. Government Printing Office, August 1969), p. 117.

As a basis for this comparison, the last column in the table lists for each facility the percentage of the Dresden 1 discharge rate. The percentages for the other boiling water reactors (BWRs) are comparable with Dresden 1. Gaseous discharges from the pressurized water reactors (PWRs) are several orders of magnitude below the BWRs. The main reason for this difference is that PWRs have the ability to store gaseous wastes for a decay period of about 60 days prior to discharge, whereas BWRs discharge short-lived gases after only about 30 minutes. The relative factors adjusted for equal generating capacity for the PWRs range from about 20 times (Indian Point discharge and soluble fly ash) to about 2,000 times (Yankee discharge and insoluble fly ash) less than the coal plant.

Discussion of Comparisons No attempts were made in these comparisons to determine the effects of the particle sizes of fly ash on the exposure dose; it was assumed that all the airborne natural radioactivity collected was in particles of respirable size. Measurements of particle sizes in filters from Widows Creek indicate that all particles were below 10 microns. Solubility fractions for the radioactivity also were not determined for fly ash, a factor that has an appreciable effect on the dose produced in humans. The exposure dose rates were, however, calculated for bone and lung (soluble and insoluble) to determine the range of this effect; the fraction of the ICRP limit for the bone dose was a factor of 2.5 higher than the lung dose fraction.

The Widows Creek data, upon which this study was based, are believed to be sufficient to support the results of the comparisons. Results of air sampling at the Turkey Point site and the H. B. Robinson site indicated that special methods under carefully selected conditions were required to measure airborne fly ash radioactivity from which exposure dose could be calculated. The Widows Creek site satisfied the selection requirements: high discharges of coal fly ash at a site with poor dispersion characteristics. It was located in a valley; the stacks were short; and there was channeling of the plumes in the valley. High sample volumes and negligible effects from background dust assured good measurements of the plume concentrations. The comparisons were based conservatively on the two highest concentrations which were measured at Stations 1 and 5. Results from the five other stations supported this comparison by defining the plume and indicating that the general variation of concentration within the downwind area was as expected.

Meteorological predictions indicated that the maximum concentration from Units 1 through 6 would be about 4 miles downwind.[8] The highest radium-226 concentration was measured at 1.7 miles, the closest point of measurement downwind. This observation suggests that a building-wake effect occurred and that concentrations at points closer to the stack could have been higher. Comparison on this basis with Dresden measurements, which were made at 1.1 miles, would be reduced somewhat but not enough to reverse the relative conclusion. The Dresden site, on the other hand, is located on flat land and does not have valley channeling or wake effects.

A comparison of nuclear and fossil plants from a health standpoint should consider the radiation exposure significance of liquid radioactive wastes from nuclear plants and the health effects of chemical pollutants from fossil plants. These comparisons were not made because of the difficulty presented by having no relatable health-effect index. The presence of liquid radioactive wastes obviously adds some increment to the exposure dose from reactors but from a different pathway than the submersion dose rates that were considered for airborne releases.

Comparison of New Power Plants
A comparison of new coal plants with modern BWRs and PWRs was made in order to determine the relative radiological significance of the various plants utilizing current technology. The

coal plants are about 1,000 MW in size and have efficient air-cleaning equipment and tall stacks.[11] The results of this comparison are shown in Table 9.5. The PWR data in the table are based on the Connecticut Yankee Power Plant,[13] which is one of the newest PWR plants in operation at present. The BWR data are based on Dresden 1, and the listed dose rates are those obtained during field measurements adjusted to the annual average discharge rate reported for 1968.[10] The measured dose rates are within a factor of 2 of those predicted by meteorological calculations. Dresden 1 is assumed to be typical of BWRs of current design although no operating data are available for newer BWRs to confirm this assumption.

Table 9.5
Comparison of Modern Nuclear Plants with a Modern Coal-Fired Plant

	Coal plant[a]	PWR[b]	BWR[c]
Size	1,000 MW	462 MW	200 MW
Stack	800 ft	200 ft	300 ft
Stack (effective)	1,500 ft	0 ft	300 ft
Stack discharges			
Fly ash	4.5×10^9 g/yr	—	—
Radium and thorium	47.9 mCi/yr		
Noble gases	—	3.7 Ci/yr	240,000 Ci/yr
Liquid discharges			
Fission products	—	3.8 Ci/yr	6.0 Ci/yr
Tritium	—	1,735 Ci/yr	2.9 Ci/yr
Critical organ (ICRP)	Bone	Total body	Total body
Dose limit (ICRP)	333 μrem/h	57 μrem/h	57 μrem/h
Dose rate	35.2×10^{-6} μrem/h-MW	1.2×10^{-6} μrem/h-MW	8.7×10^{-2} μrem/h-MW
Fraction ICRP dose/MW	10.6×10^{-8}	2.1×10^{-8}	1.53×10^{-3}

[a] Coal plant data are based on 9% ash content, 97.5% fly ash removal, and radioactivity levels in Widows Creek coal.[12]
[b] PWR data for the Connecticut Yankee plant are based on 1968 discharge data.[10]
[c] BWR data for Dresden 1; dose rates were linearly adjusted from field measurements to 1968 discharge data.[10]

The coal plant data in Table 9.5 represent a typical plant burning coal with 9% ash content that contains the same natural radioactivity as Widows Creek coal, has 97.5% air-cleaning efficiency, and an 800-foot stack with heated plume rise to about 1,500 feet. The dose rates presented in Table 9.5 were calculated as fractions of ICRP limits per megawatt. On this basis, the dose from the coal plant is 5 times greater than the PWR, but about 14,000 times less than the BWR. For insoluble fly ash, fractions of ICRP lung dose rates for the coal plant are about twice as large as ICRP fractions for the PWR and about 35,000 times less than those for the BWR.

The relative improvement of the new coal plants with respect to the new nuclear plants is attributed to increased air-cleaning efficiency and higher stacks that provide much greater dilution of the effluent before it reaches ground level. These stacks, however, diffuse the effluent over a larger area and thus could result in exposures to more people at a lower level. Such a practice is undesirable, since the highest population dose commitment (man-rems) is related to the total risk that must be assumed.[9] Even though these stacks may allow a meeting of standards, the resulting reduction of individual radiation exposure dose rates may be offset by the increased impact on a larger population. In order to account for the variation of the comparison due to high stacks, the hypothetical coal plant in Table 9.5 was assumed to discharge its radioactivity from an effective stack height of 300 feet, which is the same as the BWR stack. No change was made in the PWR release conditions, which assume a ground-level release. The results of this comparison are shown in Table 9.6. When compared on this basis, the coal plant, as a fraction of ICRP recommendations, is about 400 times greater than the modern PWR or about 180 times less than the BWR. The factors are only relative, however, and actual exposures are of negligible public health significance.

If long-term storage of gaseous wastes is incorporated into

Table 9.6
Comparison of Modern Nuclear Plants with Typical New Coal Plant Normalized to a 300-Foot Effective Stack Discharge Height

	Coal plant[a]	PWR[b]	BWR[c]
Stack (effective)	300 ft	0 ft	300 ft
Radium and thorium	47.9 mCi/yr	nil	nil
Noble gases	nil	3.7 Ci/yr	240,000 Ci/yr
Fraction ICRP dose/MW	8.6×10^{-6}	2.1×10^{-8}	1.53×10^{-3}

[a] Coal plant data are based on 9% ash content, 97.5% fly ash removal, and radioactivity levels in Widows Creek coal.[12]
[b] PWR data for the Connecticut Yankee plant are based on 1968 discharge data.[10]
[c] BWR data for Dresden 1; dose rates were linearly adjusted from field measurements to 1968 discharge data.[10]

future BWR designs, the results of all of these comparisons will change. The design of the Shoreham Nuclear Power Station,[14] for example, indicates that levels following the usual 30-minute storage will be reduced by an additional factor of 60 if stored for nine days.

Long-Term Effects
Both nuclear and fossil-fuel plants will influence the radiation exposure doses received from radioactivity in the world's atmosphere. One way in which fossil-fuel plants influence radiation doses is by the "Suess Effect." [15] Carbon-14 exists in nature and represents a dose rate to man of between 1.0 and 1.6 millirem per year. Prior to 1900, the carbon-14 distribution within the carbon cycle was in a "steady state." Since 1900, however, the combustion of coal and oil has added to the atmosphere an enormous amount of CO_2 that is free of carbon-14, thereby reducing the specific activity of carbon-14 in the atmosphere and in those reservoirs in rapid exchange with it. This phenomenon, the Suess Effect, will reduce the dose to mankind between 1954 and 2000 by about 3 millirem, assuming that forecasts of electricity generated by fossil fuels are correct.[15]

The most significant long-term effect from generating electricity by nuclear power is the introduction of krypton-85 into the atmosphere. Discharge of krypton-85 at operating nuclear power plants is low since it is retained in the fuel. Larger amounts are released, however, when the spent fuel is reprocessed. Because of its half-life of 10.3 years, krypton-85 is building up in the atmosphere as the amount of nuclear power generation increases. Coleman and Liberace[16] have estimated that the dose from krypton-85 in the atmosphere will be about 2 millirads per year in 2000 and about 50 millirads per year in 2060. Comparison with fossil-fuel plants is difficult. Fossil-fuel plants use a fuel that contains natural radioactivity; they do not create radioactivity as nuclear plants do. The radioactivity discharged from fossil-fuel plants is distributed locally, sometimes increasing specific activity of soil, sometimes decreasing it. The effects of nuclear plants and associated fuel-reprocessing facilities, however, are global and will be evident for very long periods of time. The exposure doses will increase unless effective removal processes are developed for krypton-85 in the gaseous discharges of fuel-reprocessing plants. Such processes are under developemnt, but it is not known when they will be applicable to fuel-reprocessing technology.

Conclusions
This study, on the basis of measurements and calculations at oil, coal, and nuclear power stations, has produced the following conclusions on the relative radiological significance of fossil fuel and nuclear power plants:
1.
A comparison of the radiological significance of fossil-fuel and nuclear power plants is most difficult because gaseous radionuclides from nuclear plants are predominantly noble gases that produce whole body exposure, whereas radionuclides in fly ash, if soluble, are long-lived bone seekers and, if insoluble, lead to radiation exposure of the lungs.

2.
Coal plants discharge considerably more natural radioactivity than oil plants of equivalent size and design. Measurements of radioactivity levels in environmental media at the Turkey Point site show no detectable change in these media due to discharges of oil fly ash. Measurements at the Widows Creek plant, an older, less efficient coal station with short stacks and low-efficiency air-cleaning equipment, showed detectable radioactivity in air only.

3.
The measured or estimated dose rates were calculated as fractions of the respective ICRP recommended limits, and these fractions were used for comparing the plants. When these fractions were related on a per-megawatt-electrical basis, it was found that the noble gases from a boiling water reactor can produce more radiation exposure than the natural radioactivity emitted from an older coal plant and that this coal plant can produce more radiation exposure than noble gases from a pressurized water reactor.

4.
The largest variations in the comparisons were caused by the storage time for gaseous wastes at nuclear plants, and by the efficiency of air cleaning and stack height for fossil plants.

5.
For power plants of current design, nuclear power reactors represent a greater overall radiological burden on the environment than fossil-fuel plants if both short-term and long-term effects are considered.

6.
If the exposure dose rates determined in these comparisons are appropriately extrapolated by annual meteorological variations, it is found that fossil-fuel and nuclear power plants are well below radiation protection guidelines recommended by the Federal Radiation Council.

7.
The radiological exposure for fossil-fuel plants is of negligible

public health significance and, for this reason, further study of the environmental aspects of fossil-fuel plants should focus on other pollutants.

References

1
J. E. Martin, E. D. Harward, and D. T. Oakley, "Comparison of radioactivity from fossil fuel and nuclear power plants," in *Environmental Effects of Producing Electric Power, Part I*, Hearings before the Joint Committee on Atomic Energy, 91st Congress, 1st session (Washington: U.S. Government Printing Office, 1969), Appendix 14, pp. 773–809.

2
R. J. Bayliss and H. M. Whaite, "A study of radium alpha-activity of coal ash, and particulate emission at a Sydney power station," a paper presented at the 1965 Clean Air Conference, University of New South Wales, Australia.

3
G. F. Stone, "A hazards evaluation of radioactivity in fly ash discharged from TVA steam plants" (Tennessee Valley Authority, Industrial and Air Hygiene Branch, February 1967), unpublished report.

4
M. Eisenbud and H. G. Petrow, "Radioactivity in the atmospheric effluents of power plants that use fossil fuels," *Science, 144*, 288–289 (1964).

5
J. A. Gordon, *Interim Report of the Study of Public Health Aspects of Fossil-Fuel and Nuclear Power Plants*, Southeastern Radiological Health Laboratory (Montgomery, Alabama: U.S. Public Health Service, Bureau of Radiological Health, August 1968).

6
Bureau of Radiological Health, *Radiological Surveillance*

Studies at a Boiling Water Reactor, Report DER-69-2 (Rockville, Maryland: U.S. Public Health Service, October 1969).

7

Bureau of Radiological Health, *Environmental Effects of Fossil-Fuel and Nuclear Power Plants — Progress Report No. 1* (Rockville, Maryland: U.S. Public Health Service, October 1968).

8

J. M. Smith, *Description of the Widows Creek Steam Plant,* Bureau of Radiological Health (Rockville, Maryland: U.S. Public Health Service, June 1969).

9

The International Commission on Radiological Protection, *Recommendations of ICRP on Permissible Dose for Internal Radiation,* Report of Committee II (London: Pergamon Press, 1959).

10

Joint Committee on Atomic Energy, 91st Congress, 1st session, *Selected Materials on Environmental Effects of Producing Electric Power* (Washington: U.S. Government Printing Office, August 1969), pp. 114–119.

11

The Energy Policy Staff, Office of Science and Technology, *Considerations Affecting Steam Power Plant Site Selection* (Washington: U.S. Government Printing Office, December 1968); also in reference 10, pp. 145–283.

12

J. G. Terrill, E. D. Harward, and I. P. Leggett, Jr., "Environmental aspects of nuclear and conventional power plants," *Industrial Medicine and Surgery, 36,* No. 6, 412–419 (July 1967).

13

U.S. Atomic Energy Commission, *Connecticut Yankee Nuclear Power Station, Final Safety Analysis Report,* Docket No. 50-213, AEC Public Document Room, Washington, D.C. The AEC Public Document Room is open to the public. It is located at

1717 H Street NW, Washington. Inquiries can be addressed to the Atomic Energy Commission, Washington, D.C. 20545.

14
U.S. Atomic Energy Commission, *Shoreham Nuclear Power Station, Preliminary Safety Analysis Report,* Docket No. 50-322, AEC Public Document Room, Washington, D.C.

15
U.N. General Assembly, 17th session, *Report of the United Nations Scientific Committee on the Effects of Atomic Radiation,* Supplement No. 16 (A/5216) (New York: United Nations, 1962), pp. 217, 249.

16
J. R. Coleman and R. Liberace, "Nuclear power production and estimated krypton-85 levels," *Radiological Health Data and Reports, 7,* No. 11, 615–621 (November 1966).

Discussion Questions for Part II

Question from Chauncey Starr, Dean of the School of Engineering and Applied Science at the University of California, Los Angeles, and formerly President of Atomics International.

Considering the comparative public health consequences of alternative energy supplies, what would be the effect on the choice of central station power plants of accepting radiation levels that are ten times lower than at present?

Answer by Merril Eisenbud.

The central power stations could operate at one-tenth the present levels with essentially no changes in design. However, more careful operating procedures would be required, since there would be that much less margin for error in regard to the quantities of radioactive wastes that could be outboarded in a given situation.

I seriously doubt that it would be necessary to reduce the emissions from present levels if the population dose were limited to 0.017 rem instead of 0.17 rem. As long as the maximum exposure to any given individual is limited to 0.5 rem, the per capita population dose will be below 0.017 rem.

Question from William A. Thomas, who is in the Ecological Sciences Division of the Oak Ridge National Laboratory.

Should states be allowed to establish environmental quality standards more stringent than those set by the federal government?

Answer by Merril Eisenbud.

I am in favor of uniform standards established by the federal government. I see no reason why residents of one state should be exposed to more or less pollution than residents in another state.

Question from Earl J. Bell, Associate Professor of Urban Planning at the University of Washington, and Chairman of the Washington Environmental Council Thermal Power Plant Committee.

If nuclear reactors are demonstrably safe, why does the AEC not permit them to be located within metropolitan areas of concentrated population, and why should the Price-Anderson Act, which absolves a utility of all liability and transfers the liability to the public, not be repealed?
Answer by Walter Belter.

Up to now, the AEC has not approved the location of nuclear power plants close to cities. AEC regulations require that a judgment be made during the regulatory process for a nuclear power plant as to the proper balance between engineered safety features and other means of assuring safety, including distance from population centers. To date, the AEC has not licensed large nuclear power plants to be located in or near large population centers. AEC testimony before the Joint Committee on Atomic Energy in 1967 pointed out that urban siting required further important advances in reactor plant design, particularly in the capability of safety systems and engineered safety features, and in adopting these systems so that they can be inspected and tested. Until these data and further research and development results are obtained and more experience is gained in the design, construction, and operation of nuclear plants, the AEC plans to maintain a conservative approach in evaluating plant safety and in establishing a balance between compensating engineering safety features and population density.

It is the judgment of the AEC and Congress that the probability of a catastrophic nuclear accident is extremely small but cannot be ruled out as absolutely zero. It is because of this remote, but nevertheless possible, chance of an accident occurring with losses that would exceed the maximum available commercial insurance that the Price-Anderson indemnity system was established by Congress.

Under the Price-Anderson Act, indemnity protection is provided to licensees and all others who may be liable for up to $500 million over and above the nuclear liability insurance or other forms of financial protection that the commission requires licensees to maintain. Licensees of large power reactors are required to maintain financial protection equal to the total amount of nuclear insurance available, currently $82 million. In the case of large power reactors, the total protection of indemnity and underlying financial protection that is available is $560 million.

The provision of the act establishing the limit of liability of persons indemnified (that is, persons legally liable for the accident) is consistent with the congressional purpose of removing the deterrent to private industrial participation in the atomic energy program posed by the threat, however remote, of tremendous potential liability claims exceeding privately available insurance. Having provided an unprecedented amount of indemnity protection — $500 million — in addition to the underlying private insurance, it can reasonably be assumed that Congress considered that adequate provision had been made to cover public liability for a reactor accident, and therefore a finite limit of liability on persons indemnified was appropriate.

Question from David MacLean, Associate Professor of Chemistry at Gustavus Adolphus College in Minnesota.

What measures are being taken to insure that the less technologically proficient countries are minimizing emissions from their nuclear installations? What will prevent those countries from using shortcuts on safety and environmental protection to minimize time and capital outlay on nuclear power generation? How does the AEC assure itself that environmental and safety condi-

tions will be met in the export of nuclear technology to underdeveloped countries?
Answer by Walter Belter.

The International Atomic Energy Agency, comprising 103 member states, requires in Article III of its statute:

The Agency is authorized to establish or adopt, in collaboration with the competent organs of the United Nations and with specialized agencies concerned, standards of safety for protection of health and minimizing danger to life and property and to provide for the application of these standards to its own operations as well as to the operations making use of materials, services, equipment, facilities and information made available by the Agency or at its request or under its control or supervision and to provide for the application of these standards, at the request of the parties, to operations under any bilateral or multilateral arrangement, or at the request of a member state, to any of the states activities in the field of Atomic Energy.

The agency's health and safety measures, approved by the board of governors on March 31, 1960, provide that agency safety standards specify maximum permissible levels of exposure to radiation and fundamental operational principles and that the safety standards are based on the recommendations of the International Commission on Radiological Protection.

The agency believes that the limits established in the basic Safety Standards for Radiation Protection (IAEA Safety Series No. 9), based as far as possible on the recommendations of the ICRP, provide an appropriate regulatory basis for the protection of the health and safety of employees and the public without imposing undue burdens upon users of radioactive materials.

The AEC assures itself that the environmental and safety conditions will be met in the export of nuclear technology to underdeveloped countries through IAEA environmental monitoring (IAEA Safety Series No. 16) and with the assurance that, in nuclear power generation facilities built by U.S. manufacturers, they have met all U.S. domestic nuclear power regulations for health and safety. In addition, the manufacturer provides training programs for operational and regulatory personnel of the underdeveloped countries.

Question from Sidney J. Socolar, Assistant Professor of Physiology at Columbia University, and Glenn L. Paulson, graduate fellow at The Rockefeller University and member of the New York Scientists' Committee for Public Information.

Comparisons have often been made between the average natural background exposure to radiation and the standards for exposure of the population as a whole to added amounts of man-induced radiation. It seems that an assumption is made that exposure to these natural levels is safe. Is there any specific information available, particularly from epidemiological studies, on the biological effects of the natural levels of radioactivity found in different areas of the world?

Answer by Merril Eisenbud.

Extensive physical measurements have been undertaken in many parts of the world, including Brazil, India, midwestern United States, and New England. While attempts to obtain epidemiological information have been made, the efforts have failed because of the lack of adequate population size and the difficulty in obtaining good retrospective medical information.

For example, leukemia occurs seven times a year in a population of 100,000 people, on the average. There are, of course, statistical deviations from the average because of unknown ethnic and environmental factors. It has been estimated that to reveal a detectable increase in leukemia at a dose of 0.17 rem per year, it would be necessary to accumulate 6 million people-years of exposure in the age bracket of 35–44 years. It has not been possible to find populations of adequate size with medical records of suitable quality.

III Hydroelectric Power

10

Ecological Effects of   Karl F. Lagler
Hydroelectric Dams

Electric power generation is one of the primary purposes for which dams have been and are being built in increasing numbers on rivers the world over. Looking to the future, one can foresee that all of the major rivers of the world will be dammed, along with many minor ones. Most often the man-made lakes created by these dams serve many purposes other than water storage for electric power generation. Nevertheless these purposes, as well as downstream water uses, remain subservient to the hydrologic management imposed by hydroelectric power generation. Both the dam and the imposed water-management system are often incompatible with regional ecology. Experience has shown that many undesirable impacts can be overcome, circumvented, or even turned to human advantage if planning is adequate and if the additional costs of social and ecological adjustment are included in the costs of the scheme. If they are, they must finally be passed on to the consumer of the electric energy or met by subsidy.

Few actions of which man is capable in a similar time and place can have such far-reaching ecological effects — physically, biologically, and socioeconomically — as a major hydroelectric dam. Many of these effects are dynamically irreversible, at least until such time as the dam is removed or abandoned for its primary use. Thus, in many of their aspects, dams and their reservoirs are potentially agents of environmental degradation, be they constructed for hydroelectricity, for one of their other recognized primary uses, or for multipurpose use. (See Figure 10.1.)

The ecological impacts of dammed reservoirs are similar, regardless of their location or the primary purpose for which they are so often solely planned and cost-benefited. These impacts

The references for Chapter 10 are on pages 154–157.

10.1
The Fontana Dam in the Great Smoky Mountains. Aesthetically pleasing and functionally efficient, a hydroelectric dam may nevertheless have far-reaching ecological effects. Photo courtesy of Tennessee Valley Authority.

are highly complex, have ecologically strong aspects, both positive and negative,[1] and remain in need of thorough, systematic analysis. The construction of models of dam impacts could make them better understood in order that future planning and costing may help to minimize ecologically dangerous and destructive aspects — even to the extent of turning some of them to gain.

Although waterpower from large dams continues ostensibly to be the cheapest source of electric energy in many parts of the world, it is often cheap only because of the failure of governments fully to reckon the secondary social and ecological costs. On a global scale such costs may indeed be exceeding the benefits of cheap electricity, especially when one considers the advancing technologies in thermal, nuclear, and small-dam alternatives for power generation that may ultimately eliminate the undesirable ecological impacts of large dams. Secondary costs of hydrodams take the form of environmental alteration, destruction, or obliteration and involve human, water, land, and living terrestrial and aquatic resources.

Planning and construction of large hydroelectric dams in the world has accelerated during the past twenty years and goes on ignoring known concepts of total ecosystems and of established principles of integrated river-basin or regional development.[2] The ecologist and social scientist are still not participating adequately in the early thinking stages regarding proposed dams for electrical power generation. The same of course may be noted for dams built for other primary purposes, such as irrigation, navigation, and flood control. It might almost be said that ecologically a dam is a dam even though its size and location are overriding factors that may minimize its stresses on the environment. One must not overlook that for some parts of the world, the primary motivation for dam construction is not economic but social — one-upsmanship and national pride — and that for such dams special criteria must be created to evaluate their effects; certainly for these installations the strategy for ecosystem development as outlined by E. P. Odum[3] is not used.

From a long-range ecological point of view, damming rivers to form lakes or reservoirs may be regarded as a human counterforce to the inexorable process of ecological succession by which nature exterminates lakes through filling. This counterforce, for example, in the United States, exclusive of Alaska and the Great Lakes, has produced more square miles of man-made reservoirs than there are of natural lakes.[4] Thus, were it not for numerous catastrophic ecological effects of shorter range, man's environmental conscience could be pretty clear regarding damming. These catastrophic effects are amplified when flooding of land takes place in heavily populated river valleys, especially in the tropics, where there are extra hazards of aquatic vectors of disease, where living is more prevalently at the subsistence level, and where public health education is likely to be sorely limited.

The damming of rivers for the creation of reservoirs of water still has an undetermined role in the world's critical water balance. Less understood is the global role of reducing the runoff of fresh water to the sea as a form of water conservation to make it available as a coolant to prevent overheating of the planet.[5]

Physical Systems

A simplified representation (Figure 10.2) of related ecological, social, and economic effects of a hydroelectric dam would show a first impact on the hydrologic system with possible implications for the earth's crustal system and the atmospheric system. The major effect of an altered hydrologic system is first to be seen in the terrestrial ecosystem and the aquatic ecosystem. Ultimately, the events in the terrestrial and aquatic ecosystems are reflected in the production, communication, and power systems. The first three of the foregoing systems are physical; the second two, biological; and the last three, human.

Although all of the associated systems are in some ways disrupted by a dam, the effects of most common concern arise in the aquatic and terrestrial ecosytems and their relationships to the human systems of production and communication.

Hydrologic System Effects The first hydrologic effects of a dam on a river are evidenced above the dam when the river becomes a lake and the physically unstable riverine ecosystem shifts into a relatively much more stable lacustrine ecosystem. This may be preceded downstream by an even more sudden and catastrophic

10 Ecological Effects of Hydroelectric Dams

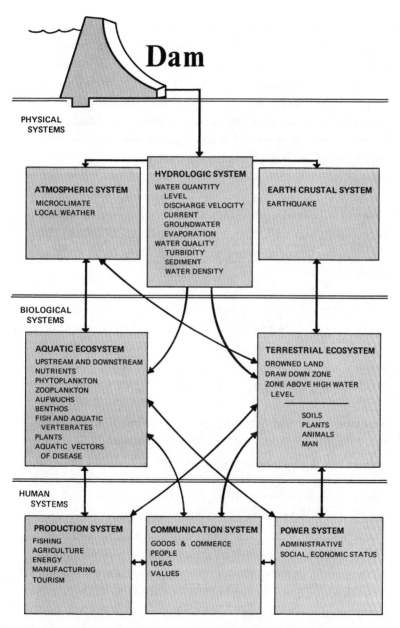

10.2
A simplified representation of related ecological, social, and economic effects of a hydroelectric dam. Adapted from a presentation by Gilbert F. White.

change if the river is temporarily eliminated, as is sometimes done by arresting the discharge completely until the reservoir behind the dam has filled for the first time. For an already rare component of the flora and fauna, such an event can be disastrous.

In general, the changes in the hydrologic system that may result from construction and operation of a hydroelectric dam are physical but directly or indirectly bring about changes in all of the dependent biological and human systems (Figure 10.2). These changes may include some or all of the following:

1.
Relative stabilization of water level in the basin with resultant effects on volume of discharge and current velocity downstream — all affecting the energy-flow regime in the living parts of the total ecosystem.

2.
Increased input to ground water supply, with possible distant benefits (as hoped for by percolation from Lake Nasser* into Nubian sandstone of the region with useful outputs for agricultural and domestic water supply in distant depressions of the desert and in existing small oases).

3.
Reduction in turbidity through settling-basin action (which can mean locally intensified sedimentation) and possibly also through reduction of erosion in the new lake in contrast to the previous riverine situation. Further probable reduction of turbidity downstream may also reflect settling-basin action of the reservoir in addition to benefits of stabilized water flow through the system.

4.
Increase in basin evaporation loss due to perennial existence of an expanded open area of water surface and perhaps accentuated by evapotranspiration of emergent aquatic plants (if they develop extensively).

*Lake Nasser is formed by the new high dam on the Nile River near Aswan, United Arab Republic.

5.
Alteration of aquatic temperature regime not only within the reservoir (where diurnal fluctuations will be reduced from those of previous riverine conditions and where seasonal thermal stratification may develop) but also downstream, where thermal characteristics will change under the influence of the lake-water outflow through the dam, particularly in relation to the level in the dam from which the water is drawn. Depending on the level of the penstock openings in the dam, the temperature (as well as the water chemistry) may be affected for various extents downstream. For example, in reservoirs of the Temperate Zone, where water is drawn from below the thermocline, growth and survival of cold-water fishes may be augmented[6] in the tailwater, but warm-water species may be disfavored. Such tailwater effects may extend for miles downstream. They can be adjusted as desired if multilevel discharge ports are built into the dam[7] and if spillage through them is feasible in the water/power budget. (See Figures 10.3 and 10.4.)

6.
Change in water chemistry is detectable within the reservoir where chemostratification may be represented by deep-water oxygenless zones (perhaps also laden with hydrogen sulfide) — zones that cannot support fishlife and which may be reflected downstream for variously limited reaches that are also uninhabitable to fish (again depending on the level at which water is drawn through the dam for power generation). Within the reservoir, the decomposition of submerged vegetative and organic soil components may produce an explosive release of chemical nutrients to the biosystem.

Atmospheric System Effects Depending on a multiplicity of factors including moisture content, temperature, and movement of air masses, along with regional topography, compass orientation, and size of the reservoir, local microclimate and even gross

10.3
The Norris Dam. Water chemistry may be affected by the level of penstock openings. Spillports low in a dam may draw stagnant, unproductive water from the depths of the reservoir, quickly aerate it, and restore it to productivity in the tailwater below the dam. Photo courtesy of Tennessee Valley Authority.

10.4
The Kentucky Dam on the Tennessee River. A new fishery in the tailwater may result below a hydroelectric dam. Photo courtesy of Tennessee Valley Authority.

weather may be changed by a hydroelectric impoundment. Although this change may be small from man's point of view, the chain of events among other animals (for example, insects) or among plants may have strong secondary elements of human concern.

Earth Crustal System Effects Dams and man-made lakes may have seismic effects,[8] but knowledge is still meager on this subject. Nevertheless it is possible that dam failure and oftime resultant catastrophic flooding may stem from an unanticipated earth crustal reaction.

Biological Systems
The biological systems in the environs of, and downstream from, a hydroelectric reservoir usually show sharp changes as a result

of the dam effects on the hydrologic system. The biological system is composed of interrelated terrestrial and aquatic ecosystems in which the affected components are spatial aspects of the environments, the substrates, solar energy harnessing and transfer mechanisms, and human environment. The affected biological system in turn has effects on the production, communication, and power components of the human system and feedback to the hydrological system.

Terrestrial Ecosystem Effects   In the terrestrial ecosystem there are many environmental changes that most often result from damming:

1.
The terrestrial habitat shrinks above the dam as land is drowned by filling of the new reservoir, but the land–water interface increases. Both these factors will be reflected in floral and faunal changes, the appearance of new dominants and population complexes, and spontaneous as well as dictated shifts in the distribution of the human population. Often one of the most dramatic effects is witnessed in the sepulchral forest of submerged trees, the skeletons of many of which remain visible above the surface of the reservoir (See Figure 10.5).

2.
A more or less distinctively new habitat is formed in the drawdown zone that is alternately and seasonally inundated and exposed. In this zone, and in land that borders it, soil moisture increases even to the extent of forming localized marshes or swamps, depending on the terrain. Again, there are accompanying floral and faunal changes.

3.
Downstream, if seasonal flooding has been arrested by the dam, long-established patterns of water–soil-fertility relationships may be altered — usually with net reduction of soil moisture content and changes in nutrient input and nutrient cycling. Floral and

10 Ecological Effects of Hydroelectric Dams

10.5
Volta Lake, Ghana. In drowning a forest, a hydroelectric reservoir may not only have effects on the plants and animals; it may impede navigational uses. Photo courtesy of J. McAllister.

faunal changes ensue, including stabilization of opportunities for human habitation along with other aspects of the human system — especially agricultural and industrial production and all-weather land communication. There may even be measurable changes for hundreds of miles downstream where seasonal saturation of soil and inputs of nutrients are arrested (as in the Nile Valley, below the new high dam near Aswan).

Aquatic Ecosystem Effects   In the aquatic ecosystem there are likewise many far-reaching changes that result from damming. These changes differ somewhat upstream from those downstream of the dam and center on the following components: aquatic nutrients, plankton, aufwuchs, benthos, and higher flora and fauna (including fishery organisms and aquatic nuisance vectors).

1.
Initial flooding that drowns plants, animals, and organic soil components sets the stage for a sudden release of nutrients into the water. The process of this release is oxidative and at least initially places an elevated biochemical oxygen demand on the system. Often, as a result, stagnant limnetic regions of the reservoir become both devoid of oxygen and laden with hydrogen sulfide. Either condition eliminates possibilities for useful aquatic production in parts of the system. The foregoing, however, is a relatively unimportant phenomenon in comparison with the overall ecological impact of the accompanying release of nutrients. Biogeochemical enrichment of the water results in explosive growths of indigenous phytoplankton and of the periphyton component of the aufwuchs. This biological explosion at the level of primary production is rapidly reflected in augmented and accelerated transfers up the various food chains, most of them leading to fishes. The aufwuchs is favored by the expanded substrate provided by trees and shrubs left standing in the shoal regions of the reservoir.[9] It is also favored by developing beds of higher aquatic plants.

2.
The same initially explosive nutrient release commonly effects an upsurge in the density and extent of higher aquatic plants. This upsurge may ultimately lead to serious conflicts with various aspects of the human system,[10] such as fishing, movement of boats, and even with the mechanics of power generation if turbines or water intakes for them become clogged. Overabundant growth of water weeds also hastens filling of the reservoir basin and may

eliminate parts of it from useful aquatic production. Dense beds of such plants may also favor a rise in abundance of aquatic disease vectors such as the schistosome snail hosts of swimmer's itch or of the dread bilharziasis. Quiet water in beds of aquatic plants favors the survival of mosquito larvae, including the vector species of malaria. Through evapotranspiration, emergent aquatic plants may greatly accentuate normal evaporation losses from the surface of the reservoir.

3.

As a reservoir begins to fill with water, there is a transition from riverine to lacustrine ecology; rheophilic habitats disappear and limnetic ones expand. Naturally, floral and faunal changes reflect these habitat shifts. Organisms that are current-dependent decline or disappear whereas quiet-water organisms become dominant. While becoming dominant, the latter may also exhibit amazing rates of growth and survival due both to the expanding environment and to the accompanying initial nutrient release, uptake, and transfer, as described earlier.

4.

Instability is a characteristic of the aquatic populations in new reservoirs. This instability is both quantitative and qualitative and reflects the species composition of the populations at their beginnings in the lake as well as the impacts of the changes in the hydrologic, terrestrial, and human systems.[10] Contributing to the instability may also be the introduced elements of flora and fauna; a man-made lake seems to have an almost certain appeal for human tampering in the form of introduction of exotics.

5.

Downstream changes are manifested in many of the components of the aquatic ecosystem that are also changed in the lake that develops upstream from a hydroelectric dam — but in different degrees, for different distances, and sometimes in different directions.

The most dramatic effects downstream result from stabilization of stream flow as flood stages are eliminated or greatly reduced. When floods are passed downstream at all, their timing is altered because of retention by the reservoir, and their character and effects are consequently different from floods of historic schedule. Extremes of low water may also be eliminated; no longer may the stream go "dry-to-pools" in the dry season, with water movement confined within the substrate. In addition to reduction in variation of discharge, the substrate itself is usually stabilized, destructive scouring and erosion may be minimized, and turbidity reduced. Each of these hydrologic factors has biological effects typically reflected in changes in the numbers and distribution of species populations and sometimes in changes in the species composition of the flora and fauna.

Survival of sedentary and attached organisms, plant and animal, is favored by stabilized stream flow. Aufwuchs and rooted aquatic plants may increase as may populations of rheophilic organisms in general. Unfortunately, in some parts of the world certain of the favored rheophils are nuisances, such as the blackflies *(Simuliidae)* collectively, or disease vectors (such as *Simulium damnosum,* vector of river blindness, or onchocerciasis). Elimination of floods may, however, disrupt reproduction of valuable fishes that are floodwater spawners, or it may disrupt recruitment through elimination of floodwater nursery grounds for young that are dependent on the enriched and accelerated nutrient and energy cycling in the wide spread of an inundation zone. Related effects may extend to seafront, estuarine zones if seasonal or overall annual dilution of salinity is reduced by stabilized discharge or by a lessened annual discharge due to augmented evaporation losses in the reservoir system.

For migratory aquatic organisms, especially fishes, a hydroelectric dam may be a barrier that is ultimately destructive of a species population.[11] This applies to diadromous fishes of both the anadromous and catadromous kinds, and to potamodromous species as well. For upstream migrants, the barrier effect may be reduced by the development and installation of fish passes and the operation of fish hatcheries,[11] although the many fish passes constructed to date have been far from universally successful.[12] Stocking of fish for introduction or for maintenance is not always successful either. But for species that migrate downstream, either as young or as adults, doom may lurk if passage is forced to take place through the hydroelectric turbines.

Human Systems
The human system is affected both directly and indirectly by a dam that creates a water-storage reservoir. Not the least among these effects are those of social change with all of its many and inevitable ecological overtones.

Direct effects are those arising from inundation of existing segments of the production, communication, and power components of the human system. Indirect effects are those that result from interactions of the hydrologic system and force secondary changes in the human system complex.

The initial effect of a dam in a populated area is psychological stress. Uncertainty seems to be the earliest factor of this stress. It arises from imperfections in planning and in knowledge relative to date of first filling and delineation of the future shoreline. These imperfections are compounded by faults in the dissemination of information and in the creation of understanding among affected people relative to the whole scheme. In some parts of the world some such people have never seen a lake or an electric light. Psychological stress may result in the loss of human power through reduction of efficiency in the production system. For example, inputs in agriculture may be wholly or partly interrupted and repairs and maintenance of property, roads, and trails abandoned as people wait to see what will happen.

As the water begins to rise behind the dam, human stress may begin to peak when flooding obliterates homes, land sources of subsistence or income, community platforms, landmarks, communication routes, burial and otherwise hallowed grounds, historical and archeological sites — all that is familiar. The stress really peaks for many people when relocation begins. Most often the move is to unfamiliar ground, in some situations already inhabited by a host population with its own established social and economic organizations. The host population may also evidence stress depending on how the affairs for transfer of people among them has been handled, including preparation and monetary adjustment and carrying capacity of the land. Host stress may range from little or mere acceptance to bitter antagonism. Some areas that have been chosen for resettlement have been unable to support the population under available conditions for agriculture, necessitating further relocation.[2]

The stress on resettled people and their likelihood to commit ecological rape during the period of readjustment are minimized if they are permitted to participate in the plans for their removal, as well as in decisions as to where and how they will be relocated. An additional assist is found in their participation in the working out of indemnification schedules for property and crops lost, for stress, and for reestablishment at new sites. Where the entire move is by fiat and fraught with misunderstanding aggravated by poor preparation and communication, stress is likely to be so severe that organized resistance may develop.

Following loss of the familiar and faced with the unfamiliar, stress is likely to continue for some years as adjustment is made to a new location and, often, to a new way of life. However, under an optimal scheme where the relocated and host people have had an opportunity to participate at all stages of planning, the result can be improvement in the environmental outlook of the region which will inevitably yield improvements in the biological system and in the production and communication systems as well. It is nevertheless recognized that not all human responses to the changed ecology can be planned. For example, at Volta Lake in Ghana, where some 80,000 people were relocated to communities and housing of an improved character, many of these people plus others (totaling perhaps 100,000 by the end of 1969) had migrated spontaneously to the shore of the new lake and had established unplanned settlements. The ecological effects of such spurious migration can be very serious throughout both the biological and human systems of the new reservoir and its environs.

Production System Effects The production system may undergo rapid changes even before formation of the reservoir behind the dam. Such a change arises with the organization of a work crew and with preparations for the import of heavy equipment and construction materials. Often experts and laborers are also imported, and villages, which may become permanent, are created for them. Planning for such accommodations has seldom been done in ways compatible with the local ecological situation. Local workers remaining after the surge of construction is past may be forced to live off the land, with consequent dire effects on the flora and fauna.

Where the rise in incidence of water-associated diseases is a part of the ecological cost of a hydroelectric dam and reservoir, measurable losses may result in human power for production.[13]

After formation of the new reservoir, new kinds of agriculture and livestock production may emerge, both with long-range ecological effects. The potential for subsistence, artisanal, commercial, and recreational fisheries increases over that of the former river and is less likely to be so seasonal. The attraction of water and its multiple recreational uses can bring significant local impacts from a tourist industry.

The energy available from a hydroelectric plant may not only provide new amenities but may lead to local industrialization. Industrial development may have deleterious ecological effects unless properly planned and integrated with environmental capabilities for tolerance and regeneration. Problems of pollution will emerge.

Regional population growth is the handmaiden of a hydroelectric dam and reservoir in many parts of the world. Such growth is also classifiable as a secondary ecological effect, which in turn serves to accelerate yet other changes. Along with it, problems of domestic water supply and waste disposal will arise. There will also be new needs for planning in regional development.

Along with the foregoing expansions in the production systems, commerce is likely to witness a boom. Goods, supplies, and services will not only increase in demand but many of them will be new, demanding increased investments in training and in business capital. In some parts of the world, there will be an accompanying shift from barter to cash economy.[10] Where fishing and fishery industries arise in an area previously populated by farmers, retraining may not only be mandatory but difficult.

Communication System Effects   The communication system at the site of a dam and reservoir is initially severely strained, then fractured, and subsequently is often only haphazardly restored. Happily the restoration may bring needed improvements with it. The initial strain felt on the communication system is for the movement of engineering supplies and equipment for dam construction. These may result quickly in local improvements of roads, for example, to provide access to the dam site. Unless carefully planned, the cultural development along these roads can be ecologically and aesthetically very destructive.

Flooding behind the dam drowns both the formal and the informal routes of travel, cutting regular arteries of communication. Sections of roads and trails may be eliminated and river bridges and fords impassably inundated. In a typical pattern, riverine craft become the first vehicles for water transport. Most often such craft are not seaworthy for traversing the open reaches of water that appear. While organized water transport is being

10 Ecological Effects of Hydroelectric Dams

10.6 Kpandu, Ghana, on the shore of Volta Lake. When a reservoir cuts a road, a boat-docking area may develop spuriously and encourage the spread of water-vectored diseases. Photo courtesy of J. McAllister.

planned, spontaneous forms develop, often also with limited safety qualifications. Commonly unplanned settlements arise at the cut ends of roads and, in addition to showing little respect for habitat preservation, provide local sites of intensified risk for waterborne or water-associated diseases (See Figure 10.6).

The temporary breakdown of transportation has impacts on the movement of both goods and people. Impacts are also witnessed on the transfer and exchange of ideas, information, and values. Restoration of a communication system may, however, bring improvements over previous conditions. Water transport may be more economical than the previous land transport. Watchfulness is required, nevertheless, where new roads are constructed to make sure that construction procedures are ecologically compatible. New road cuts, for example, are often serious sources of erosion and siltation.

Power System Effects   In the human system, the power structure reflects impacts of a hydroelectric dam. Because of the services required for adjustments to the altered biological, production, and communication systems, power is more likely to become centralized in the agency, governmental or corporate, responsible for the dam or hydroelectric production. Changes in the power structure inevitably have long-term social and land-use effects.

Conclusion
Environmental changes arising from a hydrodam and reservoir are many and far-reaching in their effects. Their interactions are still so little known that ecological scientists remain in a poor position to influence entrepreneurs, engineers, economists, and government leaders caught in the whirl of development. Working knowledge in the ecology of man-made lakes is so scant that even where planning precedes construction by fifteen years, many things go wrong and other things happen that were unforeseen.

Such was the situation, especially with reference to resettlement, for the hydroelectric dam at Akosombo that made Volta Lake in Ghana.[14] A possibly worse set of problems is arising in connection with the construction of Kossou Dam, which will impound Bandama Lake in the Republic of the Ivory Coast. Here the whirl brought the secondary ecological problems into focus only when the construction of the dam had begun in 1969.[15]

Encouraging are three recent international symposia on hydrodams and lakes and the publications emanating therefrom: one in England in 1965,[16] one in Ghana in 1966,[17] and the third scheduled for Knoxville, Tennessee, in 1971 under the leadership of the Scientific Committee on Water Research (COWAR) of the International Council of Scientific Unions. Also encouraging is a recent series of experience papers on such schemes in Africa[18] and bibliographical efforts.[19,20]

Further encouragement lies in assistance given to governments for research and development by the United Nations Development Program with the Food and Agriculture Organization as the executive agency.[21] There have also been spinoffs from these projects into related private reservoir research, some in connection with the International Biological Program. Long-term integrated planning continues to precede major construction efforts in the Mekong River Basin of Southeast Asia.[22]

In conclusion, the negative and positive ecological aspects of each hydroelectric dam and reservoir differ. People troubled by man's dam building and concomitant landscape scarring wrongly assume that all of the ecological consequences must be bad.[23] Negative or positive responses also differ from the vantage point of the critic or the defender — meaning one thing to an economic developer, industrialist, sport or subsistence fisherman, water skier, or water viewer and another thing to a man whose home is drowned, whose form of livelihood is interrupted, whose urine turns to blood, whose eyes go blind, or who shakes with malaria because a lake was created. Great inputs of money and technology are still required to maximize social and economic benefits of hydrodams and reservoirs skillfully and humanely and to minimize or eliminate their undesirable environmental effects. It is encouraging to note that ecological considerations had an important role in stopping the construction of at least one dam in recent times, the proposed Rampart Dam on the Yukon.[24]

In spite of the variety of alternatives available for the generation of electricity — hydro-, mechano-, thermal, or nuclear — we still need greatly to improve the application of each of these means in order to achieve environmental compatibility. Unfortunately, time seems to be running out for us to make these improvements because of the geometrically increasing demands for electric power. Perhaps what we really need is a better mouse trap. Witness the public conscience in the headline: "Power — yes; power plants — no." [25]

Regarding hydrodams and reservoirs, the need remains to recognize and be guided by the observation of Darling and Dasmann: "This planet can only be rescued for humanity from the damages of the human presence if viewed in ecological terms: that human society and nature make up a single ecosystem, and all human activities must be appraised and managed in the light of their effects on all the other components of the ecosystem." [26]

References

1
Henry van der Schalie, "Egypt's new high dam — asset or liability?" *Biologist, 42,* Nos. 3–4, 63–70 (1960).

2
Thayer Scudder, "Kariba dam: the ecological hazards of making a lake," *Natural History, 78,* No. 2, 68–72 (February 1969).

3
Eugene P. Odum, "The strategy of ecosystem development," *Science, 164,* 262–269 (1969).

4
Bureau of Sport Fisheries and Wildlife, *Sport Fishing — Today and Tomorrow,* ORRRC Study Report 7, Report to the Outdoor Recreation Resources Review Commission (Washington: U.S. Government Printing Office, 1962), 127 pp.

5
G. P. Kalinin and V. D. Bykov, "The world's water resources, present and future," *The Impact of Science on Society, 19,* No. 2, 135–150 (1969).

6
Milan Peňáz, "Das Wachstum einiger Fischarten im Svratka-Fluss und seine Anderungen unter dem Einfluss der Talsperre Vir," *Acta Scientiarum Naturalium Academiae Scientiarum Bohemoslovacae,* Brno, II n.s., No. 11 (1968), 50 pp.

7
James P. Carter, "Temperature control of reservoir releases into Nolin and Barren tailwaters," *Kentucky Fisheries Bulletin, 49,* (1968), 28 pp.

8
J.P. Rothe, "Fill a lake, start an earthquake," *New Scientist, 39,* No. 605, 75–78 (1968).

9
T. Petr, "Problems of assessment of periphyton production in a tropical man-made lake," in *Report of the Regional Meeting of Hydrobiologists in Tropical Africa* (Nairobi: UNESCO Regional Centre for Science and Technology for Africa, 1969), pp. 144–145.

10
G. W. Coulter, "What's happening at Kariba?" *New Scientist, 36,* No. 577, 750–752 (1967).

11
P. F. Elson, "Threat of industrialization to Canada's Atlantic salmon," *Fisheries of Canada, 22,* No. 5, 3–9 (1969).

12
Charles H. Clay, *Design of Fishways and Other Fish Facilities,* (Ottawa: Canada Department of Fisheries, 1961), 301 pp.
13
Carl E. Taylor and Marie-François Hall, "Health, population, and economic development," *Science, 157,* 651–657 (1967).
14
James Moxon, *Volta: Man's Greatest Lake* (London: Andre Deutsch, 1969), 256 pp.
15
Pierre Fallet, "Kossou: 85,000 paysans à la recherche de terres nouvelles," *Eburnea,* No. 24, 2–5 (April 1969).
16
R. H. Lowe-McConnell, ed., *Man-Made Lakes* (New York: Academic Press, 1966), 218 pp.
17
Letitia E. Obeng, ed., *Man-Made Lakes: The Accra Symposium* (London: Oxford University Press, 1969), 398 pp.
18
Neville Rubin and William M. Warren, eds., *Dams in Africa* (London: Frank Cass and Company, 1968), 188 pp.
19
Gabrielle Edgcomb, *Man-Made Lakes: A Selected Guide to the Literature,* prepared for the Africa Science Board in cooperation with the African Section of the Library of Congress (Washington: National Academy of Sciences, September 1965), 97 pp.
20
Robert M. Jenkins, *Bibliography on Reservoir Fishery Biology in North America,* Research Report 68 (Washington: U.S. Bureau of Sport Fisheries and Wildlife, 1965), 57 pp.
21
Karl F. Lagler, ed., *Man-Made Lakes Planning and Development* (Rome: Food and Agriculture Organization of the United Nations, 1969), 71 pp., 33 figs.

22
Committee for Coordination of Investigations of the Lower Mekong Basin, *Semi-Annual Report: 1 January–30 June 1969*, Vol. I, 29 pp., and Vol. II, 189 pp. (Bangkok: United Nations Economic Commission for Asia and the Far East, 1969).

23
Robert M. Jenkins, *Reservoir Fishery Research Strategy and Tactics,* Circular 196 (Washington: U.S. Bureau of Sport Fisheries and Wildlife, 1964), 12 pp.

24
Stephen H. Spurr et al., *Rampart Dam and the Economic Development of Alaska, Vol. I,* Summary Report (Ann Arbor: University of Michigan, School of Natural Resources, 1966), 61 pp.

25
Resources for the Future, "Power — yes; power plants — no," *Resources,* No. 33, 4–6 (1970).

26
Frank Fraser Darling and Raymond F. Dasmann, "The ecosystem view of human society," *The Impact of Science on Society, 19,* No. 2, 109–121 (1969).

11

Pumped Storage   David A. Berkowitz
Hydroelectric Projects

The demand for electric power consists of a base-load requirement that is steady throughout the day and a peak-load requirement that is superimposed on the base load. The peak-load requirement varies on an hourly, daily, weekly, or seasonal basis, depending on the character of the local power market. In order to meet peak demands, power companies have several alternatives. Installed base-load capacity can be great enough to supply peak demands, in which case the power plant will be operating at less than optimum efficiency for much of the time. In certain locations, it might be possible to purchase power from a distant location where peak demands appear at a different time during the day. Finally, the power company can use generating equipment specifically relegated to delivery of the peak requirement. Conventional hydroelectric dams with limited reservoir capacity are frequently employed for meeting peak loads. It is also possible to install gas turbine, diesel, or quick-cycle steam-generating units. Another approach to meeting peak loads is to store energy that could be generated through utilization of off-peak generating capacity in a form that could be readily changed back into electrical power when the demand peak occurs. A practical way to accomplish energy storage is to use available, off-peak capacity to raise water to a higher level. Power companies have been turning more and more to pumped storage hydroelectric stations in order to meet peak-load requirements.

Technical Description

The power required to lift 1,000 gallons of water 1,000 feet in one minute is approximately 190 kilowatts, which results in slightly more than 3 kilowatt-hours of energy stored in the

Editors' note. This paper was received by the editors on April 6, 1970. The references for Chapter 11 are on pages 171–172.

11 Pumped Storage Hydroelectric Projects 159

raised water. This energy could be released at any later time when the water is permitted to work its way downhill again. In practical pumped storage projects, there are frictional energy losses as water passes through tunnels and penstocks, and pumping and generation inefficiencies. Typically, 1.5 kilowatt-hours of pumping energy are required for each kilowatt-hour of energy generated subsequently.[1]

The present tendency in pumped storage hydroelectric units is toward reversible units that pump in one direction and generate in the other. On a single shaft, a pump–turbine-motor–generator unit would pump water up to an elevated basin or storage reservoir in one direction of rotation. Water is returned along the same path, causing the pump–turbine-motor–generator unit to rotate in the other direction and generate electricity. (The powerhouse and penstock leading to the upper reservoir at the Yards Creek Pumped Storage Generating Station are illustrated in Figure 11.1.) The reversible units appear to offer simpler construction and equipment requirements.[2] Other types of pumped storage installations employ separate pump and generator units. And in this case, pump and generator units need not be located at the same station. For example, the pump could be located at upstream reaches of a watercourse where the pumping head to the upper reservoir would be smaller than the head of water at the location of the generator.

Installed generating capacity of conventional hydroelectric plants in January 1968 was 46 million kilowatts, which represented approximately 18% of the nation's total generation capacity. Additional facilities planned and under construction by the federal government or licensed to nonfederal owners by the Federal Power Commission would approximately double the present installed capacity. Total hydroelectric potential of the United States has been estimated as 175 million kilowatts, exclusive of pumped storage.[1] We are nearing saturation in the exploitation of conventional hydroelectric power in the United

11.1
Powerhouse of the Yards Creek Pumped Storage Generating Station in Blairstown, New Jersey. This is a pure pumped storage project that operates between two reservoirs. The lower one is seen in the foreground; the upper is on top of Kittatinny Mountain, 760 feet higher. The penstock leading into the powerhouse is 19 feet in diameter. There are three pump/turbine units, each with a generating capacity of 110,000 kilowatts. A site restoration program is under way which includes soil stabilization and reforestation. Plans for additional expansion in connection with the adjacent Tocks Island development on the Delaware River have been opposed by environmentalists. (The station is jointly owned by Jersey Central Power and Light Company and Public Service Electric and Gas Company.)

States; over half of the estimated potential capacity has been accounted for. Remaining sites are not necessarily the most advantageous ones because of stream-flow characteristics or poor location. Furthermore, there is increasing competition for the land by other users in the industrial, commercial, recreational, and conservation sectors.

In January 1968, generating capacity of reversible pumped storage power plants in operation in the United States was 2.3 million kilowatts; capacity of 4.8 million kilowatts is under construction at the present time, and approval for capacity of over 10 million kilowatts is pending before the Federal Power Commission.[3] Newer units are of larger and larger capacity. For example, the Northfield Mountain Pumped Storage Project of Northeast Utilities Service Company, being constructed in Massachusetts, will have an initial capacity of 1 million kilowatts.[4] The Cornwall project of Consolidated Edison Company of New York at Storm King above the Hudson River will have an initial capacity of 2 million kilowatts. (Construction of the Cornwall project has been delayed pending evaluation of environmental and aesthetic factors.)

Use of pumped storage is growing in the rest of the world, as well as in United States. Japan has a capacity of 2.6 million kilowatts in operation or construction; the countries of Western Europe have 2.3 million kilowatts; and there are other projects in Brazil, Colombia, Canada, and the USSR.[2]

There are many advantages to pumped storage hydroelectric power from the point of view of the power industry. It increases the number of sites acceptable for construction of dams whose primary purpose is to supply peak-power needs. Relatively small stream flows can support large generating capacities because water is stored and a portion of it can be reused. The pure pumped storage plant does not require a large stream in a deep natural valley or a site that would otherwise be suitable for dam construction. A steep profile to the land with adequate space for construction of a storage basin at a high elevation is sufficient. Many more additional sites could be found for pumped storage plants that would be totally unacceptable for conventional hydroelectric exploitation.

Pumped storage plants have desirable electrical characteristics. They can be started up from a standstill to full electrical load, typically, in three minutes or less.[2] They can also be run continually at a partial generating load or "spinning in air" to provide a more rapid spinning or assured reserve capacity. When running as a pump, they constitute a rapidly interruptable load to release generating capacity for an unpredicted peak. They can be used running in either direction for correction of system power factor.[2]

In economic terms, the pumped storage power plant fits very easily into the cost analysis equation. The large generating capacity used in conjunction with a large base-load power plant permits the base-load plant to be operated at a higher plant factor. In addition, pumping during off-peak hours increases minimum base-load requirements that can effect initial sizing of the base-load plant. Installed costs are comparable with other schemes for achieving peak capacity, such as quick-cycling steam units or gas turbines. In the case of the Northfield Mountain Project, installed costs were estimated to be about 20% less.[4] Investment and operating costs at Northfield were also less, and of course the fuel (that is, the water) for pumped storage power plants is free. Developers of the Northfield project have estimated the most economically advantageous mix of pumped storage power in the New England region to be about 18% of total peak load;[4] site studies are under way for additional plant locations in New England. Other operational and planned pumped storage projects are well distributed across the country.

A secondary purpose for pumped storage projects is to facilitate the control and management of local water resources. Water is imported, as well as impounded, to meet the needs of agriculture and urban centers. In the simplest case, water can be lifted from one watershed where there is an abundant supply by some measure to another watershed where there is a need. At Northfield Mountain, it has been suggested that a portion of the water pumped into the upper reservoir be drained into Quabbin Reservoir.[4] At the time of spring flow, part of the Connecticut River, which normally flows south to Long Island Sound, will thereby be diverted into the water system of the Boston metropolitan area. The Tocks Island development on the Delaware River (adjacent to Yards Creek Pumped Storage Project in New Jersey, jointly owned by Jersey Central Power and Light Company and Public Service Gas and Electric Company) proposes to increase domestic water supplies in metropolitan areas of New Jersey as well as to provide pumped storage power.[5]

Water management in California has developed in a most complex manner to supply requirements of the San Francisco and Los Angeles metropolitan areas and to supply the agricultural requirements of the San Joaquin Valley.[6,7] As part of the continuing evolution of the California Water Project for water resources management, hydroelectric facilities will be installed where feasible. More power will be consumed by pumping than will be generated, however.[8] Oroville Dam on the Feather River, for example, will have reversible pump–turbine units and conventional turbines. Thermalito Diversion Dam is located below the Oroville Dam. It diverts water into Thermalito Forebay behind Thermalito Dam. This dam discharges to Thermalito Afterbay, which is an artificial impoundment that subsequently discharges again into the Feather River. Thermalito Dam will contain reversible and conventional units. There are two stages of pumping and generation in the Oroville-Thermalito project.

San Luis Dam, at the western perimeter of San Joaquin Valley, creates an off-stream reservoir with a capacity of 2 million acrefeet of water. It is filled by pumping from O'Neill Forebay. Water reaches the forebay by gravity from the California Aqueduct and by pumping from Delta-Mendota Canal. Power is generated when water is released from storage to the forebay.[8]

The city of Los Angeles has proposed the construction of dikes across arms of Lake Mead behind Hoover Dam and Castaic Reservoir in California. The dikes would create afterbay pools into which pumped storage plants would operate.[8]

The power industry and governmental organizations at federal, state, and local levels seem committed to use of pumped storage plants for increasing the effectiveness and efficiency of management of the nation's power resources as well as a means for deriving more power from the nation's water resources.

Environmental Aspects
The environmental aspects of pumped storage hydroelectric projects are complex and often difficult to predict or anticipate. Each project is a special case involving the local characteristics of terrain, water quality and flow patterns, fish populations, human factors, and the manner in which they are perturbed. Even at operational pumped storage projects, there have not been sufficient study, analysis, and long-term observation to describe adequately all of the effects that have taken place. At Muddy Run above Conowingo Dam on the Susquehanna River in Pennsylvania, however, ecological studies have been under way since 1966. (Muddy Run Pumped Storage Project is illustrated in Figure 11.2.) The studies cover benthos, plankton, and fishes in the entire Conowingo Reservoir, as well as the Muddy Run pumped storage reservoir, and have involved extensive experimental modeling of the water systems.[9] Conowingo Reservoir supports Peach Bottom Atomic Power Station, as well as Muddy Run Pumped Storage Plant, and the experimental model includes effects of thermal discharge in cooling water. Initial results of studies seem to indicate that the Muddy Run and Conowingo reservoirs constitute a healthy aquatic environment that supports an abundant warm-water fish population and many

11 Pumped Storage Hydroelectric Projects

11.2
Muddy Run Pumped Storage Project. The powerhouse is at the edge of the Susquehanna River, in the foreground. The storage reservoir in the upper left-hand corner is approximately 400 feet above the level of the river. There are four intake structures in the horseshoe-shaped canal feeding eight 110,000-kilowatt reversible pump/turbine combination units that function alternately as motor-driven pumps and turbine generators. The reservoir supports a healthy fish population. In addition, a 100-acre constant level lake was created as part of the project for recreational purposes. Photo courtesy of the Philadelphia Electric Company.

other species.[9] Nuisance formations of algae are not expected, nor is deterioration of dissolved oxygen content. The studies are continuing, however, and there is a need for similar investigations at other installations.

Part of the environmental interaction of pumped storage projects relates solely to their size and physical impact on man's visual experience of the countryside. There are the powerhouse, intake structures, tunnels and penstocks, an artificial lake, and power lines for transmission. The size of hydroelectric generating machinery tends to decrease as head of water increases. The powerhouse at Northfield Mountain Project will be underground.[4] But pumped storage offers the possibility of much greater exploitation of water resources; the prospect of hydroelectric structures and large transmission lines penetrating into scenic areas that heretofore were never considered as sites for hydroelectric projects is not greeted enthusiastically by some groups. Some power companies appear to be sensitive to needs for recreation and outdoor experiences and include in their proposed developments facilities for visitors, hikers, campers, fishermen, and hunters. Indeed, consideration of these factors, as well as effects on fish and wildlife, can be a prerequisite to licensing by the Federal Power Commission.[10] The visual and land-use aspect of pumped storage projects is only a small facet of the broader societal question of the interaction of man's technological prowess with man's visual environment. There is no clear policy to guarantee resolution of this conflict, and confrontations will probably surround pumped storage projects as well as other large-scale engineering enterprises for some time to come.

Pure pumped storage projects require an upper reservoir that is totally artificial or, occasionally, make use of a natural pond or basin that is artificially increased in capacity by excavation or construction of dikes. The lower reservoir can also be an artificial impoundment in which a stream is dammed or a swamp or wetland is filled, but in many cases it is an existing large

11 Pumped Storage Hydroelectric Projects

body of water such as a river or reservoir behind a dam. The higher the head of water between upper and lower water levels, the smaller total volume of water required to deliver a given number of kilowatt-hours of electricity. At Northfield Mountain, where the operating head is approximately 800 feet, about 10,000 acre-feet of water will be transferred to generate 6.5 million kilowatt-hours of electricity, which could readily be the energy delivered in a single day. At Muddy Run, the operating head is approximately 400 feet; more than two times as much water is required to deliver the same amount of energy. Thus, the upper reservoir of a pumped storage project can be many hundreds of acres in extent with large, daily changes in water level. At Muddy Run, the artificial lake has an area of 900 acres, with typical daily level fluctuations of 50 feet. At Northfield, the artificial lake has an area of 300 acres when full; the level will be lowered almost 60 feet in normal operations (100 acres = 0.156 square mile = 4.36×10^6 square feet = 4.05×10^5 square meters).

Large amounts of water are pumped back and forth as lakes are created and drained. When the Muddy Run station is operating at full generating capacity, water moves at a volume flow rate of typically 28,000 cubic feet per second into the impounded flow region of the Susquehanna River behind Conowingo Dam. Volume flow rate during pumping is typically 24,500 cfs.[9] The natural flow of the river into the reservoir during the summer months is between 10,000 cfs and 20,000 cfs.[9] At Northfield Mountain, water will be moved at a volume flow rate of 10,000 cfs in and out of a channel of the Connecticut River at a point about seven miles above Turners Falls Reservoir. The natural flow of the Connecticut River at that point for most of the year is less than 10,000 cfs. The river will tend to flow backward from reservoir to pumping intake during pumping operations. Consequently, existing current flows in river channels and patterns of current flow in reservoirs will be disrupted and subject to daily changes and fluctuation during periods of pumping or power generation.

Artificially reversed flows in river channels have been cited by fishery biologists as possible cause for concern. In those cases in which the river supports an anadromous fish population, fish could become disoriented and lost because they use current as a cue in their migratory movements.[11,12] This is a problem only during the times of migration, which are seasonal. Uprunning usually occurs in the spring, downrunning in the fall. However, the onset of migratory movements is related to river water temperature,[9] and exact periods during which stable current-flow patterns should be maintained are difficult to specify.

For nonmigratory fish species, the pumped reservoir offers additional living space. The route to the upper reservoir is through the pump–turbine unit. Fish are drawn to the intake and into the pump; then they are subjected to the pressure of a waterhead of hundreds of feet. They seem to survive this treatment; some of them could even make daily round-trip journeys. There is some loss in each direction when fish are killed in turbine blades, but it has been estimated that the effect of increased living space in the upper reservoir is such that the total increase in fish population more than compensates for loss in transit.[13] The fish population of the storage reservoir at Muddy Run is large enough that weekend fishing (when level fluctuations are reduced) has been suggested. Hitherto, there have been no proposed recreational uses for storage reservoirs, although at Muddy Run, a 100-acre constant level lake was created at one end of the reservoir that is only for recreational use. A dam for the formation of this lake was constructed at the time the main impounding dam was built.

Water quality in the upper reservoir will be the same as that of the river or lower reservoir from which it is drawn; there is frequent mixing and exchange. The upper reservoir usually has the natural soil or rocky bottom resulting from excavation or clearing. Presence of a large lake at a high elevation where there was none before can affect the local water table. Water could percolate into the ground and influence stability of the ground and nearby slopes. It could also mix into existing aquifers. If the lower source of water is polluted or brackish, the upper reservoir also will be. Thus, quality of local water sources could be adversely affected. A lined upper reservoir has been constructed at the Taum Sauk project in Missouri (Union Electric Company). It has an area of 55 acres and is approximately 800 feet above the lower reservoir on the East Fork of the Black River. It is totally artificial and is lined with concrete and asphalt. It leaks.[5]

Summary
There is a certain engineering elegance to storing energy for later use during peak-demand periods. If storage and subsequent release of energy can be accomplished with little loss and if there is sufficient difference between peak- and base-load demands, a more efficient total utilization and management of power generation facilities results. Power will be more reliable and perhaps more reasonable. The pumped storage hydroelectric plant offers a way to store energy for subsequent release and, in addition, provides a reserve supply for emergency demand situations or power failures. In those regions of the country in which there is a significant peak-power demand (typically 20% of the base-load demand), pumped storage appears to be an economically feasible alternative for meeting the demand when compared with simply enlarging base-load capacity or with generating peak power by some other method. Power companies are turning to pumped storage with increasing frequency.

The environmental effects of pumped storage hydroelectric projects do not seem to be as serious as those of conventional hydroelectric projects, though there is a need for additional and continuing environmental studies. The area of the artificially created lakes and impoundments for pumped storage are usually measured in hundreds of acres, whereas those behind conventional hydroelectric dams are frequently measured in hundreds or thousands of square miles (one square mile = 640 acres). The insult to the local surface water systems caused by the construction of a major dam can be severe and can effect total changes in the characteristics of the region (see Chapter 10). The pumped storage project need not cause such great change, however, because the projects are physically smaller (even though electrical generating capacity is comparable), and they constitute apendages on local water systems rather than interruptions. Water in the pumped storage system can be reused; natural flows are only required for makeup purposes and initial filling. Percolation from the upper reservoir into locally surrounding lands does raise questions involving land stability and water quality. These effects must be evaluated in initial site selection studies.

In those cases in which volume pumping rate is greater than volume flow rate of the river or stream passing by the intake or flowing into the lower reservoir, fluctuations in direction and magnitude of water currents can be expected. Similar effects have been observed in reservoirs behind conventional dams,[12] which can cause confusion and disruption of migratory fish species. Nonmigratory species seem to survive and multiply successfully in the upper reservoirs, however, and it has been suggested that the upper reservoirs be made available for sport fishing, much as the tailwaters have that are below dams (see Figure 10.4, page 141).

Pumped storage projects have introduced another way in which man can make an engineering use of his environment. Many more sites are suitable than for conventional dams. New lakes will be created; existing ponds, wetlands, or marshes destroyed or modified; and the landscape will be burdened with powerhouses (although these can be underground), transmission lines, and other facilities. The Federal Power Commission and individual power companies seem sensitive to environmental attitudes and frequently plan the proposed sites for their plants to include recreational facilities, which are described in promotional literature available to the public. The power plant "site package" now includes picnic areas, campsites, trails for hiking and snowmobiles, facilities for motorboats and waterskiing, and so on. The approach is one of multipurpose land use, hopefully an effective method for reconciling conflicting societal demands on our environmental resources. The land will no longer be a wilderness, however.

References
1
U.S. Federal Power Commission, *Hydroelectric Power Evaluation* (Washington: U.S. Government Printing Office, March 1968).
2
G. Dugan Johnson, "Worldwide pumped-storage projects," *Power Engineering, 72*, 58–63 (October 1968).
3
U.S. Federal Power Commission, Office of Public Information, Washington, D.C., private communication.
4
Sherman R. Knapp, "Pumped storage — the handmaiden of nuclear power," *IEEE Spectrum, 6*, 46–52 (April 1969).
5
H. W. Hunt, "Pumped storage, a major hydro power resource," *Civil Engineering, 38*, 48–53 (March 1968).
6
Frank M. Stead, "Desalting California," *Environment, 11*, No. 5 2–10 (June 1969).
7
State of California, Department of Water Resources, *The California Water Plan*, Bulletin No. 3 (Sacramento, May 1957).

8
James J. Stout and M. Frank Thomas, "Major pumped storage projects in the United States," *Combustion, 39,* 28–35 (January 1968).

9
Stanley Moyer and Edward C. Raney, "Thermal discharges from large nuclear plant," *Journal of the Sanitary Engineering Division, Proceedings of the American Society of Civil Engineers, 95,* No. SA6, 1131–1163 (1969).

10
U.S. Federal Power Commission, *Federal Power Commission Interests in Environmental Concerns Affecting the Electric Power and Natural Gas Industries* (Washington: U.S. Federal Power Commission, 1969).

11
Colton Bridges, superintendent of the Massachusetts Division of Fisheries and Game, Commonwealth of Massachusetts, private communications.

12
Robert F. Raleigh and Wesley J. Ebel, "Effect of Brownlee reservoir on migrations of anadromous salmonids," *Reservoir Fishery Resources Symposium* (Washington: American Fisheries Society, April 1967), pp. 415–443.

13
Edward C. Raney, Cornell University, private communication.

IV Fossil-Fuel Power

12

Clean Power from Coal,   Arthur M. Squires
at a Profit

In 1969 the United States power industry discharged to the atmosphere about 7×10^6 tons of sulfur in the form of SO_2. In absence of controls other than tall stacks, the discharge in 1980 will be some 18×10^6 tons.

I doubt that many persons, much beyond 1972, will regard the tall stack as a suitable means of control. Sweden believes itself harmed by SO_2 from Britain and the Ruhr.[1] Acting upon a resolution introduced by Sweden, the United Nations General Assembly has sent out the call for an international conference in 1972 to examine environmental problems including the question of SO_2. The conference can be expected to increase the already great political pressure for regulations that would compel power stations to reduce SO_2 discharges, whatever the height of their stacks.

New Jersey has issued rules that call for a fuel sulfur level below 0.3% by weight after October 1971. Unable to conceive where fuels of such low-sulfur level can be found, fuel suppliers are generally skeptical that these rules will "stick." The public's distaste for SO_2 is such as to discourage a politician who might wish to cater to fuel suppliers by rescinding the 0.3% limit. It is more probable that pressure will develop in neighboring states for authorities there to match New Jersey's goals.

When fuel suppliers and users accept the inevitability that SO_2 emissions must be sharply limited, what will be the technological response? In a word, I believe that power may well become cheaper.

It must be said at once that means now available for controlling SO_2 are expensive in the sense that their application would add significantly to the cost of electricity from coal-fired or oil-fired

The references for Chapter 12 are on pages 219–227.

plants. Environmentalists would reply that the added cost is a small price for clean air. Yet the attempt to weigh costs of control versus benefits is a premature and probably silly exercise if it is carried out before the technological community really accepts the idea that controls are imperative. After this acceptance, everyone may be pleasantly surprised at the inventions mothered by necessity.

There is a suggestive historical parallel to the situation. Before 1863, British alkali works poisoned the air with massive discharges of hydrogen chloride gas (far worse than SO_2). In 1863 Parliament passed the Alkali Act, requiring a 95% reduction in hydrogen chloride emissions. Before the act, managers of alkali works testified that water scrubbing to absorb hydrogen chloride had been tried and did not work. But soon after the act, gas-scrubbing towers were developed in which hydrogen chloride could be absorbed by water trickling downward over surfaces of internal packing. For a while, much of the hydrochloric acid thereby produced was taken to sea and dumped, but markets for the acid developed. Most significantly perhaps, the Alkali Act appears to have stimulated invention. Weldon's and Deacon's patents for chlorine gas, filed in 1866 and 1868, respectively, soon turned a profit from hydrogen chloride formerly wasted. Before these patents, chlorine had been a curiosity.

My purposes in this chapter are two:

1.
To argue that paths of technological development exist which could lead to suppression of SO_2 from coal and at the same time to a lower cost for power.

2.
To argue that a massive injection of money into coal engineering is the immediate ingredient necessary to open up these paths of development.

The argument that more money is needed for coal research is not new. So often the argument is politely met with a bland look that seems to say, "Oh yes, those poor devils who work on coal — and if we gave them more money, what would they do with it?" People regard coal as difficult, and so it is; but too many spokesmen for science repeat the accepted opinion that coal is too difficult to do much about. There are too few coal engineers, too few coal-engineering establishments, and too few speculative publications on opportunities for coal engineering for most scientifically literate persons to have much idea of what coal's true prospects might be. I shall try to make a plausible case that exciting opportunities for coal engineers do exist which can be realized if more support is forthcoming for their efforts.

History of Coal-Firing Techniques

Pulverized fuel (PF) firing is the combustion technique used today in all of the great power-generating stations based upon coal. Not the least of coal's "difficulties" is the maturity of this firing technique. It is much harder to overturn an established technique, particularly when equipment has grown to large sizes, than it is to introduce an entirely new technology. A large body of engineers and corporate executives are at home with the problems of PF firing. Large sections of several engineering societies devote their attention almost exclusively to these problems. Power engineers find it difficult to imagine that a radically different technique might serve better. Nonpower engineers are little encouraged to think about coal combustion; they are repelled by the specialized character of much of the work in progress to improve PF firing, as well as by the small interest shown by the power community in alternative procedures.

Minds are freer where equipment is still small or where greater interest is taken in the export of small plants to developing regions. England, France, and Germany are responsible for recent innovations in coal combustion, to be treated later in this chapter.

A parallel situation in the steel industry is instructive: Sweden's and Austria's steel operations are small in scale. A generation ago, these countries perfected major advances in blast-furnace practice and in oxygen-steelmaking, respectively. Many years of advocacy and study were necessary before the United States steel industry could bring itself to adopt the new techniques.

Such major innovations are unlikely to arise where the application of a mature practice has reached a large scale. Where this is so, a new procedure must be proved on the commercial scale before it can be adopted generally by businessmen mindful of their dual obligation to serve the public's expanding requirements on schedule and to preserve their own credit at the banks. A long period of development is inevitable before a large-scale test can be completed. The length of this period and the cost of the test themselves become arguments against the new procedure, almost destroying its credibility among specialists in the old procedures. Such specialists are poor witnesses to call upon for advice concerning the new, for they find it hard to imagine how what is so comfortably familiar might be displaced.

Something of the "inevitability" of a mature practice would be dispelled if more attention were paid to how the practice arose in the first place.

The historical concerns of students of coal combustion may be listed roughly in the order in which they arose:

1.
To burn coal with an acceptably small loss of carbon to smoke and ashes.
2.
To provide clean combustion gases suitable for heating materials liable to be spoiled by ashes, such as glass.
3.
To provide combustion gases that were sufficiently free of grit as not to constitute a neighborhood nuisance when discharged from a stack.
4.
To burn coal at the large rates of throughput required at plants generating electricity after about 1925.

5.
To meet increasingly higher standards for content of fly ash in stack gases.

Until about 1895, all techniques for burning solid fuel handled the fuel in a bed at rest. In some devices, the bed gravitated downward in a shaft. In others, the bed moved horizontally on a grate. Steam power engineers developed ingenious devices, the first patented by Watt himself in 1785, to feed coal continuously to a bed on a grate and to discharge ashes. Grit emissions from some of the grate devices were small, although their designers were more concerned with limiting losses of carbon.

The advantages of dealing with the coal in several steps were appreciated early. From 1735 onward, ironmakers in England used coke from beehive coke ovens. In these ovens, the heat needed to distill volatile matter from coal was provided by burning the volatile matter within the ovens. After 1800 an industry arose to supply illuminating gas obtained by heating coal indirectly in retorts. The gas industry marketed coke as a by-product, and in 1836 apparatus was introduced to derive from coke a dust-free fuel gas suitable for burning where cleanliness was desired. This "producer gas" was made by drawing air and steam through a deep bed of coke. In an ordinary shallow combustion bed, the inventory of carbon is relatively small, and the product of combustion is carbon dioxide. In a deep bed, providing a large carbon inventory, carbon monoxide arises as a secondary product. Carbon dioxide appears first, near the point where air enters the bed, but this species reacts with carbon farther along the bed to yield carbon monoxide. The net result is a "partial combustion" or "gasification" of the coke, and the consumption of air is somewhat less than one-half of the stoichiometric amount for complete combustion. Steam was added to control the temperature, reacting with coke to form hydrogen and additional carbon monoxide. Producer gas, then, consisted substantially of carbon monoxide, hydrogen, and nitrogen. Sulfur in the coke appeared in the gas as hydrogen sulfide (H_2S) and not SO_2, which appears when combustion is complete.

Often a major incentive to technical change has been the growth of demand for a commodity, making obsolete a technique capable of use only on a small scale. By the 1890s, cement manufacturers felt a need for equipment of larger capacity than the shaft kilns used hitherto. An attempt to operate a rotary kiln for cementmaking with producer gas was a failure. A kiln was operated satisfactorily with petroleum in 1895, but this fuel was then too expensive. The experience suggested, however, that a suitable flame might be sustained by injecting coal into a rotary kiln via an air blast from a nozzle. Shortly, the cement industry developed techniques for pulverizing coal and burning the coal powder. Thomas A. Edison participated in this work, attesting to its importance in late nineteenth-century technology.

In the early years of this century, the extraordinary growth of the electric industry brought a demand for ever-larger flows of steam. By about 1915, steam-power engineers realized that they would soon require steam flows larger than could be conveniently provided by a grate-firing technique. They felt an acute need for a new combustion procedure easier to scale upward in size than the existing grate-combustion devices. The experience of the cement industry was at hand; coal pulverizers, coal-conveying systems, and PF firing nozzles were available on the market; and steam-power engineers found it relatively inexpensive to undertake experiments on PF firing for raising steam. The work led to the Lakeside Station in Milwaukee. After the successful commissioning of two 20,000-kW turbines in this station in 1922, PF firing quickly became the choice for nearly all new power-station construction.

It is worth noting that engineers of the day regarded the PF boiler to be an advance from the standpoint of dust emission.[2] The larger amount of fine fly ash from PF combustion was deemed less undesirable than the lesser amount of grit from a grate furnace.

12 Clean Power from Coal, at a Profit

The power engineer of 1920 was more concerned for his immediate neighbors than for a city or a region. He soon heard about it if a nearby housewife found soot on her wash, but voices were not yet raised concerning insults to lung tissue by fine matter. Would PF firing have seemed attractive for development if engineers had shared something like today's concern about fly ash?

Although PF boilers have now reached giant sizes, they are basically quite simple. The firing compartment is a large rectangular box having walls entirely composed of vertical tubing. Each tube abuts its neighbor and is filled with boiling water. Pulverized coal and air are injected into the box via a number of burners. Hot combustion gases pass from the top of the chamber into a series of heat exchangers, each comprising a close array of parallel tubes that generally serve to superheat steam. A significant portion of the coal's ash is carried overhead with the combustion gases. A dry-bottom furnace, having steeply sloping walls at the bottom, allows about 80% of the ash to leave with the gases, while the remainder drops out of the bottom in solid form. A wet-bottom furnace has a relatively flatter bottom and retains ash on the bottom for a much longer time, so that about one-half of the ash leaves the furnace as a molten slag. A cyclone furnace uses a coarser grind of coal and burns the coal in an intense combustion zone in which the coal and gases whirl in cyclonic fashion. The effect is to separate about 70% to 90% of the coal's ash as a molten slag that can be tapped from the bottom. The changing attitude of the power engineer toward dust emissions is illustrated by the claim, advanced when the cyclone furnace was introduced in the 1930s, that it substantially solved the emission problem.

Although simple, the firing chamber must be huge, not so much to allow sufficient combustion volume as to provide sufficient heat-transfer surface to the boiling water. Gases laden with fly ash must be cooled within the firing chamber to a temperature, generally around 1100° to 1200°C, above which the fly ash would stick to superheater tubing and foul it irreversibly. The chamber must be designed and operated so that the maximum heat flux at any single point is safely below a critical flux that would burn out a steaming tube. The average flux is generally less than one-third of the critical maximum. At the average flux, the designer of a large power station must provide a veritable cathedral of flame. The boiler of the 1,000-MW Ravenswood Station of the Consolidated Edison Company of New York (Con Edison), seen in cross section in Figure 12.1, has two firing chambers, each about 34 by 64 feet in plan and 138 feet in height. Nevertheless, if air pollution is left out of account, the simplicity of the PF boiler is such that the reluctance of the steam engineer to give it up, especially in the larger sizes, is easily understood.

Pulverized Fuel Combustion in Trouble
Rising standards of air quality are putting the PF boiler under severe economic pressures. When the Ravenswood Station Unit

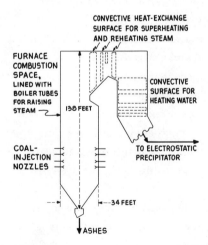

12.1
Cross-sectional view of 1,000-megawatt pulverized fuel (PF) boiler of Con Edison's Ravenswood Station. The boiler has two firing chambers, each 34 by 64 feet in plan and 138 feet in height.

12 Clean Power from Coal, at a Profit

No. 3 was authorized, Con Edison decided that the precipitator would have to provide a collection efficiency of at least 99.2%. This precipitator, seen in cross section in Figure 12.2a, is even bigger than the furnaces: it is 58 by 230 feet in plan and comprises two superposed sections, each 75 feet in height. Figure 12.2b is a photograph of the precipitator during construction. The unit cost $10 million — that is, $10 per kilowatt of the 1,000-MW capacity it handles. It has provided a collection efficiency of 99.5% in tests.

Figure 12.3 shows how dust collector sizes in Con Edison's system have increased over the years.[3] The Ravenswood precipitator operates at 370°C, while earlier precipitators generally operated at around 150°C. A reason for the higher temperature, which needs a larger precipitator to achieve comparable performance, is the introduction of coals of below 1.0% sulfur into Con Edison's system. Because ash from low-sulfur coal displays a high electrical resistivity at about 150°C, a precipitator to operate with this coal at 150°C would have to be much larger than a precipitator for a high-sulfur coal in any case, as Figure 12.4 illustrates.[3]

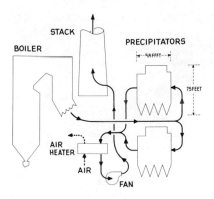

12.2a
Schematic view of 1,000-megawatt boiler and electrostatic precipitators of Con Edison's Ravenswood Station. Each of the two banks of precipitators is 58 by 230 feet in plan and 75 feet in height.

12.2b
The giant electrostatic precipitator at the Ravenswood Station, as it appeared while under construction. It was completed in 1967 at a cost of $10 million. The precipitator is larger than the furnaces it serves. It has operated at a collection efficiency of 99.5%. Photo courtesy of the Consolidated Edison Company of New York.

12 Clean Power from Coal, at a Profit

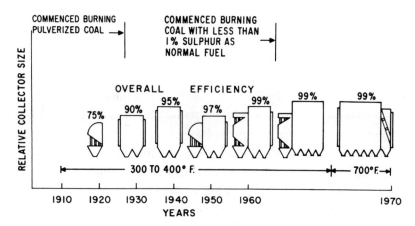

12.3
Increase in size of dust-collecting equipment for a 250-megawatt unit in the Con Edison system, after Ramsdell and Soutar.[3]

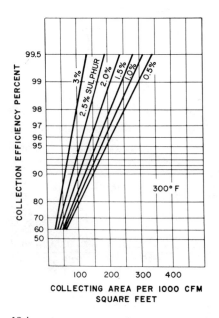

12.4
Precipitator design curves. Relationship between collecting area and collection efficiency for operation of an electrostatic precipitator at 150°C, as a function of the sulfur content of coal, after Ramsdell and Soutar.[3]

Figure 12.5 shows how costs for dust-collecting equipment in Con Edison's system have risen over the years, paralleling the increase in dust-collection efficiency.[3]

Few existing coal-fired stations are equipped with precipitators of such high efficiency as the Ravenswood precipitator. In future PF stations, the power industry will find it hard to escape a cost on the order of that incurred at Ravenswood for fly ash control.

A major drawback of PF firing for the future lies in the fact that a simple, one-step combustion places the coal's sulfur promptly into a form that is difficult to collect and recover. For typical coals, the combustion gases contain about 0.2% to 0.3% SO_2 by volume. The Ravenswood precipitator handles 4.3×10^6 cubic feet of gas

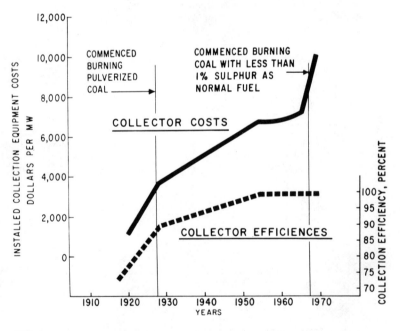

12.5
Increase in cost of dust-collecting equipment in the Con Edison system, after Ramsdell and Soutar.[3] The figures are corrected to present day costs, and include material and erection costs of collector, flues, support steel, and fly ash conveying facilities.

12 Clean Power from Coal, at a Profit

per minute, and the chemical treatment of such a vast throughput of gas for removal of a constituent present in such small amounts is almost certain to lead to apparatus of high cost.

Since the 1930s, research and development teams have worked upon many ingenious ideas for capturing SO_2 in stack gases from PF boilers. The history of these efforts is depressingly similar — initial enthusiasm followed by depression and abandonment when the economic facts became clear. So far as I am aware, only one system is now being offered to American power companies backed by responsible guarantees of performance. Even this system has yet to pass the test of operation on a commercial scale, and it cannot yet be considered commercially available in the normal sense. This is the catalytic oxidation system developed by Monsanto Chemical Company and Metropolitan Edison Company. These companies deserve great credit for having put up their own money to carry this system through the pilot scale. It cannot be considered a bargain, however, and no commercial installation has been arranged for. The system has the disadvantage that it yields a sulfur by-product in form of a relatively weak sulfuric acid (about 70% H_2SO_4), which may not be readily marketable in many parts of the country. The system is being offered at a capital cost in the neighborhood of $35 per kilowatt. As a rule of thumb, at average sulfur levels of coal or residual oil burned in the United States, a capital cost of $10 per kilowatt would be satisfactory for a system that captures a sulfur by-product from stack gases. Sale of the sulfur by-product would provide income to offset a capital outlay of $10 per kilowatt, and no increase in cost of electricity need result from the application of the system. From time to time, recently with increasing frequency, announcements are made, or gossip makes the rounds, of new stack-gas cleaning ideas that would permit a close approach to the $10-per-kilowatt target; but a healthy skepticism concerning such reports is permissible until hard facts are available.

A misleading result announced in 1963[4] gave rise to a false hope that sulfur oxides could be absorbed from combustion gases simply by injecting pulverized limestone or dolomite into the PF furnace in an amount stoichiometrically equal to the sulfur to be absorbed. The installation would be simple and low in cost, although no sulfur by-product would be obtained. The idea was especially attractive as a cheap retrofit to control emissions from old equipment during an air pollution episode. In the PF furnace, the limestone or dolomite would calcine to form CaO or $[CaO + MgO]$, respectively, which would react with SO_2 and O_2 to form $CaSO_4$ or $[CaSO_4 + MgO]$.

Unfortunately, $CaSO_4$ has a larger molecular volume than $CaCO_3$ or CaO. The first $CaSO_4$ reaction product forms a shallow, impervious layer that seals off the interior of a particle of limestone or dolomite from further reaction.[5] Good sulfur removal can be achieved only by supplying limestone in a far greater than stoichiometric amount or by reducing the stone to an exceedingly fine powder. Either expedient greatly increases the cost, making the technique unattractive in large boilers. The Tennessee Valley Authority will shortly conduct field trials of the limestone-injection technique in the hope of obtaining 40%—60% sulfur oxide removal through application of twice the stoichiometric amount of stone.[6] If the trials are successful, this technique may prove valuable in relatively small existing boilers in many parts of the country. The hoped-for results would not meet New Jersey's requirements for October 1971, namely, an emission equivalent to not more than 0.3% sulfur in fuel. The technique will not seem attractive to designers of the many stations of 1,000 MW and above which will be built in the 1970s.

Combustion Engineering, Inc., in collaboration with Detroit Edison Company, demonstrated on a small scale a system in which limestone is injected into the PF furnace in substantially the stoichiometric amount, and stack gases are scrubbed with a slurry of calcined stone in water. An advantage of the system is that fly ash is removed in the scrubber as well as sulfur oxides, so that no electrostatic precipitator is necessary. Good results were obtained in the small-scale tests, but the first commercial trials at the Meramec Station of Union Electric Company ran into difficulties. A somewhat different design at Lawrence Station of Kansas Power & Light Company has given better results and has recently performed continuously for some weeks. Although the system provides no by-products, cost estimates published recently by TVA[6] suggest that the system may represent the best hope for control of sulfur oxides from PF combustion. It is to be hoped that Combustion Engineering will soon be able to offer the system backed by performance guarantees.

Rethinking Coal Combustion
Schemes to control sulfur from PF combustion have a makeshift, tacked-on aspect. The time is at hand to rethink the problem of burning coal, with air pollution as a first consideration rather than the last.

We have already remarked that PF combustion might not have seemed so attractive to the engineer of 1920 if he had been as much concerned with fly ash as with grit. Instead, he might well have concentrated upon ways to increase the burning capacity of his familiar grate devices.

An idea was at hand. Figure 12.6 is copied from the specification of Winkler's historic patent[7] for a gas generator, filed in Germany in 1922 and put into commercial practice there in 1926. Winkler's idea was to increase the rate of gas flow upward through a granular bed to and beyond the point at which each particle in the bed was buoyed by the rising gas. When the pull of gravity upon each particle was canceled by the upward drag of the current of gas, the particles flowed freely and the bed took on the character of a boiling liquid.

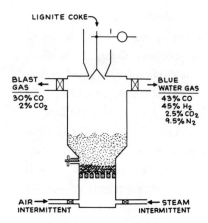

12.6
Winkler's historic idea for partial combustion of coal in a fluidized bed, from his patent application filed in 1922.[7]

Winkler's patent was germinal for major developments by the chemical engineering profession after about 1940. Novelties continue to appear, and recently there has hardly been a meeting of the American Institute of Chemical Engineers without its session on fluidization. The technique lends itself to projection to equipment of an enormous scale. Fluidized beds have been built which are scores of feet in diameter and which treat throughputs on the scale of tens of thousands of tons per day. No upper limit in the size of a single fluidized bed is foreseen.

It is hard to escape the feeling that PF combustion might never have been developed, other than for cementmaking, if Winkler's work and the chemical engineers' interest in fluidization had appeared twenty years sooner. This proposition is of course not worth arguing, but at least three concrete developments are in the offing which should dispel the aura of inevitability that has colored much recent thinking about PF combustion and the SO_2 problem:

1.
A large "Ignifluid" boiler to be built in Pennsylvania.

2.
BCURA's (formerly the British Coal Utilization Research Association) test of fluidized-bed combustion at high pressure.
3.
Lurgi's application of its historic high-pressure gas producer to power generation.

A review of these developments leads to the inescapable impression that a revolution in coal power practice is at hand. The significance of the fact that these are respectively French, British, and German developments has already been noted.

The Ignifluid Boiler Albert Godel of France had the engagingly simple idea of redesigning the wind chest of a traveling-grate furnace to allow for an air flow sufficiently great to fluidize a bed of coal resting on the grate. Figures 12.7 and 12.8 are cross-sectional views of the Ignifluid boiler, whose development grew from this idea.[8,9] Godel made the discovery that the ash of substantially all coals is self-adhering at a temperature in the vicinity of 1100°C, no matter how much higher the ASTM ash-softening temperature may be. Godel exploits this discovery to burn an extremely wide range of coals in the fluidized bed seen in Figures 12.7 and 12.8. Coal is supplied in sizes up to 3/4 inch. As a coal particle burns,

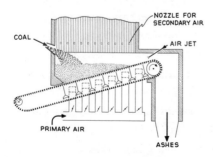

12.7
Cross-sectional view of lower part of Godel's Ignifluid boiler, showing fluidized bed resting upon a traveling grate.

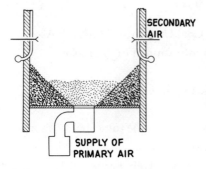

12.8
Sectional view of Godel's Ignifluid boiler across the traveling grate, showing the fluidized combustion bed between two banks of static coal.

ash is released. Ash sticks to ash and not to coal, and ash agglomerates form. They sink to the grate, which carries them to an ash pit. Godel's bed operates adiabatically, except for radiation from the upper surface. He limits the bed to the desired ash-sintering temperature (generally between about 1050° and 1200°C) by maintaining a high inventory of carbon in the bed, so that the combustion is only partial. As in producer gas, carbon is present as CO and sulfur is present as H_2S. Godel admits secondary-combustion air to the space above the bed, where CO burns to CO_2 and H_2S burns to SO_2.

As a result of his high fluidizing-gas velocity (about 10 feet per second) and low air-to-fuel ratio, the coal-treating capacity of Godel's traveling grate is roughly 10 times greater than that of previous grate-combustion devices. Godel is able to challenge the PF boiler for large power station use, and he has already supplied a boiler for 60 MW. Negotiations are well advanced for a 275-MW unit to burn and remove accumulations of anthracite wastes in northeastern Pennsylvania. The owner of the unit, UGI Corporation, will benefit from a low fuel price, between about 12¢ and 15¢ per million Btu, and the supplier of the waste, Blue Coal Corporation, will recover valuable urban land. The waste has a high ash content, and Godel's system is uniquely capable of dealing with it (see Figure 18.6, page 326).

Godel originally thought his Ignifluid system to be useful only in small boilers and for special fuels of low reactivity or high ash content. He believes he lost many years for lack of the concept that his system might go into large utility boilers. This experience illustrates the way in which the aura of inevitability surrounding a mature technique protects it from competitive ideas.

BCURA Experiment in Combustion at High Pressure Figure 12.9 depicts an experiment in fluidized-bed combustion now under way at BCURA Industrial Laboratories at Leatherhead, England.[10] The experiment illustrates another approach to fluidized combustion of coal which is receiving worldwide attention. The U.S. Office of Coal Research has sponsored large-scale trials conducted by Pope, Evans, and Robbins at Alexandria, Virginia.[11] The U.S. Bureau of Mines and groups in England and Australia are doing similar work.[12-14] The fluidized bed is operated non-

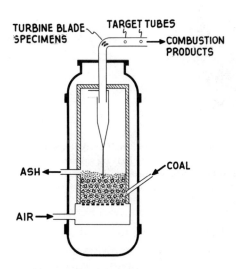

12.9a
BCURA's experimental fluidized-bed boiler operating at 6 atmospheres.

12.9b
Final assembly of the BCURA experimental fluidized bed. This boiler will treat coal at 1,000 pounds per hour. About 70% of the heating value of the coal is conveyed to steam in boiler tubes that pass horizontally through the bed; the remainder appears in the hot combustion gases as sensible heat. Photo courtesy of BCURA Industrial Laboratories.

adiabatically, the inventory of carbon is very small, and combustion is complete. The bed itself generally comprises the larger particles of the ash matter in the coal. The bed is in contact with boiler tubes that hold the temperature to a level where ash does

12 Clean Power from Coal, at a Profit

not sinter, generally below about 1000°C. Fluidizing-gas velocities of 10 to 15 feet per second have been used. The heat flux to the steam tubes is substantially uniform: the average flux and the maximum flux are essentially the same. The average flux can therefore approach much closer to the burnout flux than in a PF boiler, and less surface is needed. With the exception of BCURA's experiment at Leatherhead, the studies of fluidized combustion are being conducted at atmospheric pressure.

The fluidized bed in Figure 12.9 is housed in a pressure shell. It operates at 6 atmospheres and at the unusually low temperature of 800°C. It treats coal at 1,000 pounds per hour. About 70% of the heating value of the coal is conveyed to steam in boiler tubes that pass horizontally through the bed. The remaining heat appears as sensible heat in the hot combustion gases. BCURA's concept is that these gases would be expanded in a gas turbine. In the test, turbine-blade specimens and other metal objects are being subjected to erosive flows comparable to those that would arise in a turbine handling the combustion products. BCURA hopes the gases can be cleaned sufficiently by three cyclone stages and believes that residual ash matter in the gas will be less erosive than ash from PF combustion because no sintering will occur at the low temperature used by BCURA.

Although the test is at 6 atmospheres, BCURA proposes a pressure of 15 atmospheres for a commercial system. Figure 12.10 illustrates the dramatic reduction in boiler size which the BCURA concept should afford.[10] No substantial reduction in boiler size can be achieved by placing a PF furnace under pressure. The smaller fluidized-bed boiler should cost less. No electrostatic precipitator would be needed, and the saving might easily run over $10 per kilowatt.

Advantage of Combining Gas Turbine and Steam Power Another cost saving in the BCURA concept will arise from the fact that a gas-turbine power plant costs less than a steam plant by

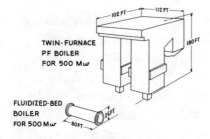

12.10
BCURA's projection of the size of a fluidized-bed boiler at 15 atmospheres, compared with a conventional pulverized fuel (PF) boiler, after Hoy and Roberts.[10]

about $30 to $50 per kilowatt. The efficiency of an open-cycle gas turbine that operates independently of a steam plant is poor, since the turbine discharges a hot gas directly to the atmosphere. Such gas turbines are ordinarily base-loaded only where fuel is cheap. Utilities have used them primarily to supply peak-load power. In a few examples, gas turbines discharge hot gases to steam boilers, and such cooperative use of gas-turbine and steam power equipment can afford base-load power at outstandingly low cost.

The gas-fired San Angelo Station of West Texas Utilities Company beautifully illustrates the economies afforded by combining gas- and steam-turbine equipment.[15] The station cost $78 per kilowatt, a remarkably low figure for its 128-MW size. In spite of its small size and modest steam conditions (99 atm and 538°C with a reheat to 538°C), the station probably has the highest efficiency of conversion of fuel to electricity in the world: about 41% on the lower heating-value basis (the fair basis for a comparison of gas- and coal-fired plants). The high efficiency results from the fact that the gas turbine "tops" the steam cycle; that is, heat rejected by the gas-turbine cycle is employed to raise steam. The proportion of the total power produced by the gas turbine is 20%, about the same as in the BCURA concept.

There is considerable excitement among boilermakers in England about the BCURA experiment. They see an early opportunity in the world export market for power stations in the several hundred megawatt class employing gas turbines of existing design.

To generate much interest among boilermakers in the United States, another hurdle must be passed. Gas turbines do not exist in a size suitable to cooperate with steam plants on the 1,000-MW and larger scale now being built here. The market for gas-turbine equipment has not encouraged development of machines of such large size. The market for large combined-cycle stations that will fire only oil or gas is far too small. The Russians have commissioned a gas turbine of 100 MW, and machines several times bigger than this can be built when means for firing coal become available.

The temperature of the gases entering the gas turbine in BCURA's concept, 800°C, is a bit low by current standards. Figure 12.11 illustrates, very approximately, the trends of temperature of gases entering aircraft and stationary gas turbines since World War II. Already stationary turbines are base-loaded

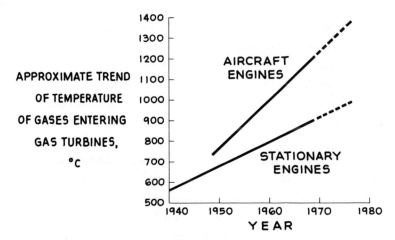

12.11
Approximate temperatures of gases entering gas turbines in aircraft and stationary applications.

at 900°C. The drive for higher performance in aircraft engines will continue, and experience from such engines will maintain the upward trend of temperatures employed for stationary machines. Indeed, only a modest extra effort would greatly shorten the delay between introduction of equipment for a given temperature into aircraft and its introduction on the ground.

As gas-turbine temperatures rise, the gas turbine of a combined-cycle system should logically provide an increasing proportion of the power. Systems can be envisaged in which the gas turbine would provide more than one-half of the power. Steam conditions even more modest than those used in the San Angelo Station might be preferred, and steam-power equipment would be low in cost. Within a decade, such systems might provide efficiencies approaching 50%[16] if gas-turbine engineers are assigned the task of achieving 1200°C, say, in stationary machines by 1980.

Incentive for Partial Combustion at High Pressure We have seen that the BCURA experiment converts 70% of the coal's energy to steam. If the gas turbine is to play a large role, more energy must be converted to sensible heat in hot combustion gases. This can be accomplished by substituting a carbon-rich bed for the carbon-lean bed of the BCURA concept, since partial combustion occurring in the carbon-rich bed can provide CO for combustion ahead of a gas turbine. Partial combustion has the incidental advantage that the quantity of gas leaving the fluidized bed is reduced by about one-half, so that a smaller bed can be specified.

A problem arises from the fact that dust carry-over from a carbon-rich bed contains a high percentage of carbon, which would represent a serious carbon loss if it is not used. Carbon in the dust carry-over cannot be consumed simply by returning the dust to the carbon-rich, partial-combustion bed. There is a tendency for the dust to be blown out of the bed again quickly, and as the dust "ages" its carbon tends to sinter to an inactive coke.

A solution to this problem is suggested by Pope, Evans, and Robbins's success in burning carbon in fines blown out of their experimental fluidized-bed boiler.[11] They provide an auxiliary bed to which air is supplied in far greater than stoichiometric amounts for combustion of the fines charged to the bed. Figure 12.12 illustrates how such a "carbon burnup" bed might cooperate with a partial-combustion bed to supply gases at high temperature to a gas turbine.

BCURA's temperature, 800°C, is probably too low for a partial-combustion bed. The reaction between CO_2 and carbon occurs at a technically acceptable rate only at temperatures above about 900° to 950°C. Both beds in Figure 12.12 might operate at about 950°C.

Carbon burnup in a partial-combustion bed becomes less of a problem if the temperature is raised into the ash-sintering range, above about 1050°C. Godel's Ignifluid unit generally pro-

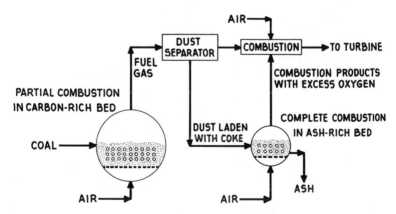

12.12
Conceptual scheme to supply CO at high pressure for combustion ahead of a gas turbine, employing partial combustion in a carbon-rich bed along with a carbon-burnup bed as demonstrated by Pope, Evans, and Robbins.[11]

vides high combustion efficiency, but even this unit runs into difficulties with refractory fuels. For such fuels, Godel provides a small current of air to the bottom of a shaft that receives agglomerated ashes from his traveling grate.

Although the present design of the Ignifluid traveling grate does not lend itself to operation at high pressure, a modification might be conceived to allow its use at high pressure. An alternative approach to partial combustion under agglomerating conditions, well suited for use at high pressure, is provided by another French development of the 1950s. Jéquier and his collaborators at Centre d'Etudes et Recherches des Charbonnages de France[17] operated a large-scale test of a unit depicted schematically in Figure 12.13. The test was at atmospheric pressure and consumed coal at rates up to a ton per hour. The bulk of the fluidized bed was carbon-rich and at a temperature between about 1000° and 1200°C. In light of Godel's experience, ash agglomerates probably formed in the bed and sank toward the grate. In the center of the grate was a cone leading to a venturi throat, beneath which an auxiliary flow of air was supplied. Within the cone was a carbon-lean zone produced by the jetting

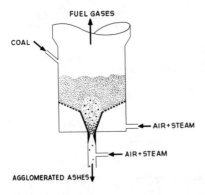

12.13
Ash-agglomerating, partial-combustion bed studied at atmospheric pressure by Jéquier and collaborators at CERCHAR.[17]

action of the flow of air from the venturi throat. This zone was at a higher temperature, about 1300°C, and further agglomeration of ash occurred therein. Spherical agglomerates of ash, about 3/4 inch in diameter, were discharged as they grew to a size big enough to fall downward against the current of air through the venturi throat. Like Godel, who retains the Ignifluid bed between banks of static coal, Jéquier designed grate and venturi to promote the formation of a static layer of solid to protect these metal parts.

Jéquier wrote in 1960: "Circumstances . . . did not allow more than short and insufficient tests to be made with this plant. Interest in the gasification of the coal fines has completely vanished with the discovery of the important natural gas sources in the South of France and in the Sahara." The last privately financed large-scale tests of coal gasification in the United States, performed by Hydrocarbon Research, Inc.,[18] and Texaco[19] in the late 1950s, were terminated for comparable reasons. Natural gas had become generally available via pipeline almost everywhere in the United States, while technological advance in converting methane to ammonia synthesis gas had made its production from coal too expensive even if the coal were free.

Jéquier's 1960 paper suggests that a carbon-rich, ash-agglomerating partial-combustion bed might be developed for operation at elevated pressure. Such a bed might advantageously take the form of the highly expanded "circulating fluid bed" whose development has been pioneered by Lurgi in Germany.[20,21] The combination of the circulating fluid bed with Jéquier's technique for development and removal of ash agglomerates is illustrated in Figure 12.14. A fluidizing-gas velocity would be selected which is about fivefold to tenfold greater than the usual fluidizing velocity for a fine powder, chosen to achieve minimum dust entrainment of the powder. The dust carry-over is then large, and a cyclone is provided to return this dust to the bed. Within the bed there is a large internal solids recirculation due to a

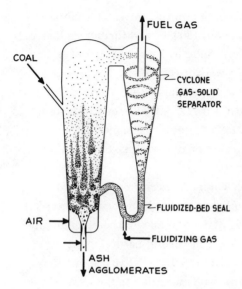

12.14
Conceptual scheme for partial combustion at high pressure, combining Jéquier's conical "valve" for ash agglomerates with Lurgi's highly expanded "circulating fluid bed."

partial demixing of the gas and solids flows. A wide range of solids residence times may be provided for, yet the solids- and gas-treating capacities can be extraordinarily high by comparison with the older fluidized-bed art. Bed temperature is stable and readily controlled.

Figure 12.14 is offered for its suggestive value and should not be taken too literally. Scaling Jéquier's conical "valve" for ash agglomerates upward in size can be expected to be troublesome. Means must be provided to deal with the net heating remaining after all CO_2 formed near the bottom of the bed has reacted with carbon to form CO. Mixing steam with the air is a poor idea, since this would correspondingly increase the latent heat loss to the stack of the power station. A recycle of products of complete combustion, containing CO_2, is a possibility. It may be possible to install a steam-raising or air-heating surface in the upper part of the circulating fluid bed or in the fluidized-bed seal between the cyclone and the bed.

In many respects, Figure 12.14 resembles the Szikla-Rozinek firing technique,[22,23] and review of experience gained in removing ash matter from Szikla-Rozinek furnaces might help in scaling Jéquier's conical valve to a large size.

We cannot leave the subject of supplying hot gases to a gas turbine without mentioning the problem of alkali salts. Figure 12.15 gives BCURA's rough estimate of the fraction of sodium in ash which is volatilized as a function of combustion temperature.[10] Turbine blades are sensitive to the content of alkali salts in the gas, and this problem becomes more serious as gas-inlet temperature is advanced. BCURA selected the temperature 800°C for its test with these facts in mind. If gases from a partial combustion conducted at a higher temperature should contain too

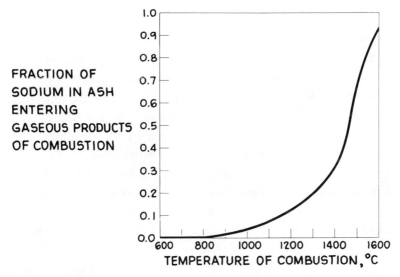

12.15
BCURA's estimate of the fraction of sodium in coal ash entering gaseous products of combustion.[10]

much alkali to be passed directly to a gas turbine, then the gases will have to be cooled and rid of alkali dust, whatever the design of the bed. Scrubbing with a heavy oil at about 370°C will be preferable to scrubbing with water at a lower temperature, so that heat may be rejected from the scrubber to a surface raising prime steam. There will be an advantage in keeping the quantity of gas to be scrubbed as small as possible. This will give the advantage to schemes in which the coal is devolatilized first, so that the partial-combustion bed must deal only with coke. This consideration, too, appears to argue against burning a major portion of the fuel in a complete combustion conducted at high pressure in a carbon-lean, ash-agglomerating bed, such as the bed studied at atmospheric pressure by Battelle.[24]

I should note the opportunity to develop a magnetohydrodynamic (MHD) device to generate electricity from the expansion of hot combustion gases with no moving parts. MHD offers a prospect of achieving 50% efficiency of power generation sooner than such an efficiency may be possible in the expected course of gas-turbine development. A balanced effort will include work on both techniques for topping the steam cycle.

Temperatures at the inlet of an MHD generator are so high as to volatilize a major part of the ash constituents of coal, if the coal is burned as such ahead of the generator. A partial combustion ahead of the generator to effect a separation of ash would simplify gas treatment for recovery of particulate matter downstream of the MHD device. This device is better suited than a gas turbine to receive gases directly from a partial combustion, since alkali presents no problem. Indeed, the potassium sometimes present in coal ash is an effective "seed" material rendering the hot gases conductive to electricity. Presence of steam in the gases is undesirable, and MHD is best suited to generate electricity from coke.

Development of systems using gas turbines or MHD will significantly reduce the discharge of waste heat from power generation.

Unit Combining Lurgi Gasifiers with Gas and Steam Turbines
Proponents of the status quo in coal-fired power generation can hardly fail to be surprised, as indeed I was, by the recent announcement that Steinkohlen-Elektrizität AG will build a combined gas-turbine, steam-turbine power plant for 170 MW in the Kellermann Power Station at Lünen, Westphalia, West Germany.[25] The gas turbine will generate 43% of the power. Clean gaseous fuel will be generated from coal by Lurgi pressure gasifiers blown with air and steam. The Lurgi gasifier is essentially a pressurized version of the apparatus introduced more than 100 years ago for making producer gas by drawing air and steam through a deep bed of fuel gravitating downward through a shaft. This gasifier has been used for more than 30 years to make town gas and is used in the world's only plant for making synthetic gasoline from coal, in South Africa. I should not have thought this gasifier to be a strong candidate for use in power generation, simply for the reason that the Lurgi needs coal in lump form. It cannot handle the fines that inevitably arise from the mining of coal and hence could not be the complete answer to gasifying coal for gas-turbine power. Also, the steam rate needed to keep ash below a clinkering temperature is so large that gas from a Lurgi unit contains more H_2 than CO, and there is a serious loss of latent heat to the stack. The largest existing Lurgi units, in South Africa, handle coal at a rate of only 300 tons (moisture- and ash-free) per day per unit, and it might prove difficult to scale the design to the far larger throughputs needed for power at the U.S. scale. It is highly significant that German engineers now regard the combination of Lurgi gasifiers with gas and steam turbines as an economic proposition on the 170-MW scale. The Kellermann Station is a signpost not to be ignored,

pointing to the need to develop gasification techniques without the Lurgi's disadvantages.

Directing Coal Combustion Developments into Paths Dealing with SO_2

Work on the BCURA concept, work on partial-combustion schemes such as those suggested in Figures 12.12 and 12.14, and work on gas turbines and MHD would be justified simply for the prospect that such work offers for reducing the cost of power from coal. A special urgency in the effort arises from the opportunity to direct these developments into paths that will lead to a combustion technology in which sulfur is dealt with early in the coal-treating process rather than at its end.

The National Air Pollution Control Administration (NAPCA) has seized this opportunity. NAPCA has supported work to explore the possibility of using limestone or dolomite in a fluidized-bed boiler to absorb SO_2 with formation of $CaSO_4$. Recently NAPCA engaged Westinghouse Electric Company to direct a broad effort toward development of nonpolluting fluidized-bed boilers. United Aircraft is exploring advanced power generation concepts for NAPCA that incorporate gas and steam turbines and coal gasification equipment.

Tests conducted so far suggest that appreciably more than the stoichiometric amount of $CaCO_3$ must be injected into a fluidized-bed boiler to capture sulfur to a degree that will meet New Jersey's goal for October 1971, namely, a maximum allowable emission equivalent to 0.3% sulfur in fuel.

If $CaCO_3$ must be used in an amount far greater than stoichiometric, the economics of the operation are improved, particularly in a station of large size, if the resulting sulfur-laden solid is treated for recovery of a valuable sulfur product by a technique that restores the solid to a form suitable for reinjection into the boiler. Consolidation Coal Company[26] has burned coal in a fluidized bed consisting substantially of particles of calcined dolomite, with good sulfur retention by the bed. The $CaSO_4$ formed was regenerated to CaO by a roast under slightly reducing conditions. Sulfur was evolved from the roast in a gas containing SO_2 at a concentration adequate for manufacture of sulfuric acid in a contact plant. British Esso[27] has conducted similar tests of combustion of residual oil in a bed of calcined dolomite and has proposed a way in which the technique might be adapted for use in boilers too small to provide a quantity of SO_2 large enough for economic production of acid. In this adaptation, the concentrated stream of SO_2 from the roast would be scrubbed with a slurry of calcined stone, thereby taking up the SO_2 in a stoichiometric amount of stone for discard.

Carbon-rich, partial-combustion fluidized beds have the advantage that sulfur as H_2S can be recovered more readily than sulfur as SO_2. It is far easier, in turn, to recover sulfur in elemental form from H_2S than from SO_2. For most locations, sulfur would be a better by-product of power generation than sulfuric acid. The former can be stored when the market is weak, and it can be shipped long distances. The latter cannot be stored readily and is expensive to ship.

Recent experiments at The City College* in New York show that the reaction to form CaS in calcined dolomite occurs homogeneously throughout the particle.[28] Figure 12.16 contrasts the kinetic situations for formation of $CaSO_4$ and CaS within calcined dolomite.

Figure 12.17 illustrates three ways in which a valuable sulfur product may be recovered from CaS. The historic Claus-Chance process, introduced around 1880, treated CaS wastes of alkali works with water and a gas containing about 40% CO_2, produced by a shaft furnace burning coke to calcine limestone. In another

---

*Work at The City College on the desulfurization of fuels by calcined dolomite is supported by Research Grant No. AP-00945 from the National Air Pollution Control Administration, Consumer Protection and Environmental Health Service.

$$[CaO+MgO] + SO_2 + \tfrac{1}{2}O_2 = [CaSO_4+MgO]$$

SHELL-PROGRESSIVE KINETICS:

$CaSO_4$ MOL. VOL. = 46.0 cm^3/g-mole

$$[CaO+MgO] + H_2S = [CaS+MgO] + H_2O$$

HOMOGENEOUS KINETICS:

CaS MOL. VOL. = 25.8 cm^3/g-mole

12.16
Contrast between kinetic mechanisms for formation of $CaSO_4$ and CaS in calcined dolomite.

step, the H_2S was converted to sulfur. Pintsch Bamag has worked on a version of the Claus-Chance method to act in cooperation with partial combustion of residual oil.[4,29]

Consolidation Coal Company,[30] British Esso,[27] and FMC Corporation[31] have worked on various procedures, each amounting to a controlled oxidation of CaS to release SO_2.

The City College is studying a technique for desorbing sulfur from CaS at high pressure by reacting the solid with steam and CO_2 to evolve H_2S.[32]

The Claus-Chance procedure has the disadvantage that heat from the reaction of CaS with water and CO_2 is wasted; also, $CaCO_3$ is recovered as a wet slime difficult to reuse except at another penalty in thermal efficiency. The oxidative procedures have the appeal of simplicity if sulfuric acid is an acceptable product, or if H_2S is available from another operation in a fuel-treating complex[31] to react with SO_2 to yield sulfur. Of the several procedures, the high-pressure desorption is probably best suited to provide a regenerated $CaCO_3$ in a form for repeated cyclic use.

Chemical species other than CaO exist, of course, which are capable of removing H_2S from a fuel gas at high temperature. Consolidation Coal has worked on the problem of providing MnO in a form suitable for this purpose,[33] and the Bureau of Mines has studied sinters containing iron oxide.[34]

12 Clean Power from Coal, at a Profit

CLAUS-CHANCE REACTION: [PINTSCH-BAMAG]

$$CaS + H_2O \text{ (LIQUID)} + CO_2 \longrightarrow CaCO_3 + H_2S$$

CONTROLLED OXIDATION: [CONSOL COAL / BRITISH ESSO / FMC CORP.]

$$CaS + \tfrac{3}{2} O_2 \longrightarrow CaO + SO_2$$

DESORPTION AT HIGH PRESSURE: [CITY COLLEGE]

$$CaS + H_2O \text{ (STEAM)} + CO_2 \longrightarrow CaCO_3 + H_2S$$

12.17
Three techniques for recovering sulfur values from CaS.

Although operation at high temperature has the advantage of better thermal efficiency, the art of scrubbing a fuel gas with an alkaline liquor to absorb H_2S is highly developed, and this alternative may be preferred in the near future.

Whatever the choice, partial combustion puts sulfur into a form easier to deal with and also provides less gas to be treated. Partial combustion at high pressure can reduce the gas volume to a small fraction of the enormous output of a PF boiler.

Long-Range Perspective for Coal

The country's and the world's reserves of hydrocarbonaceous matter largely reside in coalfields. Our own reserves of natural gas are insufficient for our growing needs, and arrangements have already been made to bring liquefied gas from abroad at costs that bring sharply into view the alternative of converting volatile matter in coal into synthetic gas. Someday even the oil of Alaska's North slope will be gone, and domestic supplies of liquid fuel will be desired.

Figure 12.18 illustrates broadly an "obvious" response to these developments. Behnke[35] has called attention to the need to study the feasibility of integrated chemical-extraction and power-producing complexes. Much of the work supported in recent years by the U.S. Office of Coal Research has been directed toward this goal.[36-41] The "Coalplex" depicted schematically in Figure

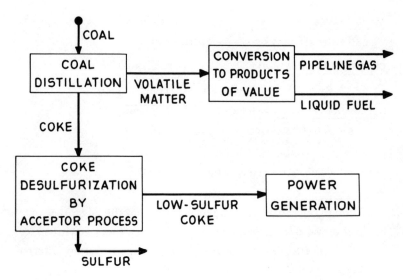

12.18
A Coalplex.

12.18 is a logical choice for study: Volatile matter in coal may be converted by relatively simple procedures into synthetic pipeline gas or liquid fuel. Fixed carbon is converted to products of value only with much more difficulty, but Consolidation Coal Company[36,42,43] has shown how easily the fixed carbon may be desulfurized through the cooperative action of hydrogen gas and a solid acceptor for sulfur in the form of H_2S, such as CaO. FMC Corporation[31,44] has operated a pilot unit demonstrating this procedure for producing a low-sulfur coke.

The term "coal distillation" in Figure 12.18 should be understood in a broad sense. A pyrolysis at high temperature, preferably under a substantial partial pressure of hydrogen, can probably lead to a Coalplex of lowest capital cost and highest overall thermal efficiency. If the product mix from such a Coalplex does not have a proper balance of gas, liquid, and electricity, the yield of liquid may be increased and that of electricity reduced by adopting more expensive procedures for the initial treatment of at least a portion of the raw coal. For example, the treatment could be conducted at lower temperature, at higher hydrogen pressure, and in the presence of a catalyst. Another alternative is to treat the coal with an aromatic solvent to extract a liquid fraction, which is then treated catalytically with hydrogen.

PF combustion is unsuited to handling the low-sulfur coke that will emerge from a Coalplex. Not only would an electrostatic precipitator be expensive (see Figure 12.4), but also the coke would need to be supplemented by a volatile fuel to maintain a stable flame. Work to develop fluidized-bed combustion techniques can be amply justified simply for the reason that these techniques will be needed to deal with low-sulfur coke certain to become available in large amounts.

A development that combines the interests of several such large industries as electricity, pipeline gas, and liquid fuel will have commercial as well as technological hurdles to overcome. The commercial hurdles may not truly disappear until natural gas is in such short supply that imported liquefied gas is no longer seen as an alternative to gas from volatile matter in coal. A less ambitious complex simply producing power and low-sulfur coke could play a useful role.

Figure 12.19 depicts broadly a logical precursor to the Coalplex of Figure 12.18.[28] This scheme would generate base-load power from the combustion of volatile matter and would ship low-sulfur coke to power stations at a distance. Heat to distill volatile matter from coal would be provided from the partial combustion of this matter, and these steps as well as coke desulfurization would occur within a single vessel housing three fluidized-bed zones. Air flow to the partial combustion is only 11% of the stoichiometric air for complete combustion of the coal, and the gases that result from the partial combustion are at high pressure (for example, 21 atmospheres). Hence, the volume

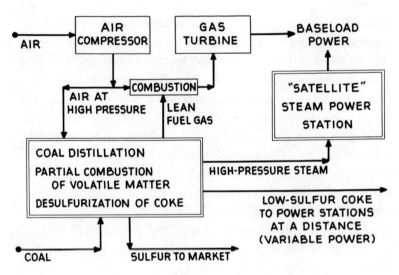

12.19
Concept for a pioneering Coalplex directed toward recovery of sulfur and generation of coal power.[28]

of gases undergoing treatment is only a tiny fraction of the volume that must be handled in a stack-gas cleaning operation. A single process vessel could treat coal for power generation at a rate of 1000 MW. The vessel would be approximately 100 feet in height, 18 feet in diameter over about 30 feet of the height, and about 10 feet in diameter elsewhere. The cost of the coal-treating equipment should be more than offset by revenues from sulfur and savings in cost of power-generating equipment (lower boiler costs, lower costs of gas turbines versus steam plant, and so on). The scheme might advantageously be installed at a riverside location to process coal on the scale of 13,000,000 tons per year, say, providing sulfur-free fuel for 5000 MW of power and typically shipping 400,000 long tons per year of sulfur. Such an installation would enjoy economies of scale in coal processing and sulfur production, and the several individual power stations that receive low-sulfur coke would be relieved of need to install equipment for these functions.

I see a natural evolution:

1.

The first Coalplexes would be justified simply for their economy in dealing with sulfur.

2.

Later, modifications would "cream off" limited amounts of pipeline gas and liquids from volatile matter. Simplicities in the processing of volatile matter would result from opportunities to throw off high-level waste heat to steam for power.

3.

As time passed, further modifications would expand production of gas or liquid.

Ultimately, in an economy powered principally by breeder reactors, a Coalplex would evolve for which power might be a relatively minor by-product, and fixed carbon would be shipped mainly for metallurgical or electrochemical use.

Reordering the Priority Accorded to Coal Engineering

Paths of technological development exist which can lead to clean and cheaper power. The missing ingredient is money. Lack of money increases the degree to which an aura of inevitability protects PF combustion from competitive ideas. When money is short, the "practical" man tends to prefer projects that aim to adjust the mature art, and it is hard to get serious attention for ideas that are not simply tacked on to the old.

I envy nuclear engineers in many respects, and not least the obvious fact that no aura of inevitability will arise to protect the light-water reactor. The breeder concept holds out the hope of an efficiency some 50 to 80 times greater than this primitive device, if efficiency is related to our total natural supplies of both fissile (uranium-235) and fertile (uranium-238 and thorium-232)

materials. The light-water reactor's aboriginality is appreciated
if it is recalled that Watt's engine had an efficiency roughly
one-tenth that of a modern power station. In a mature power
economy based upon breeder reactors, the cost of uranium or
thorium will matter very little. The light-water reactor is sensitive to the price of uranium, and so too will be the cost of initiating a breeder power economy. I have seen no responsible opinion
that the light-water reactor, at the commonly projected growth
of nuclear industry, can have any competitive standing beyond
about 1990, for there simply is not enough low-cost uranium now
in sight to fuel this reactor in the twenty-first century. It is this
fact and the uncertainties of the breeder development program
which provide powerful arguments for a reordering of priorities
for nuclear versus coal engineering. Benedict wrote recently:
"Development of the sodium-cooled fast breeder reactor will be
difficult and time consuming, and it is not certain that power
costs will be low enough to permit them to compete with plants
burning fossil fuel at today's price. Nevertheless the potential
value of having available a practically unlimited source of energy
[is] so great as fully to justify the effort now going into this development." [45] Equally justified is a vigorous effort to maintain
coal's competitive position vis-à-vis the light-water reactor, so
that some low-cost uranium-235 will remain even if the breeder is
delayed beyond present hopes.

Coal engineers need "fun money" such as nuclear engineers have
had to pursue curiosities. It is a shame that no Godel Ignifluid
unit can yet be seen here. It is a shame that no Jéquier unit has
been built here,[17] no Szikla-Rozinek unit,[22,23] no Secord slagging-grate gasifier,[46,47] no Winkler generator of the latest tuyere-blown,
pear-shaped design,[48] no Ruhrgas-Lurgi carbonization unit,[49,50]
no plant-scale version of the dilute-phase carbonization unit for
agglomerating coals developed by the Grand Forks Station of the
Bureau of Mines.[51] A theme of much of this chapter has been the

12 Clean Power from Coal, at a Profit

need to search for equipment of the highest possible coal-treating capacity. Experience with these novelties might have carried us far along a road now only dimly apparent. I have tried to indicate my view of the best paths of work to improve coal's competitive position, but it is proper to wonder how many ideas are missing for lack of the chance to see their physical embodiment.

Coal engineers also need the "spectaculars," like Shippingport, which were so important to nuclear engineers before they could offer competitive equipment. The last coal spectacular was the coal-to-oil unit at Louisiana, Missouri, on which something over $100 million was spent after World War II. President Eisenhower canceled the experiment in 1953 for what then seemed proper reasons. In some respects the plant was obsolete even when built, and by 1953 two sad facts were evident: oil from coal would not be competitive for at least two decades, and by that time a far better job could be done. Yet I wonder how many good minds were turned away from coal engineering by Louisiana's closing. Senior men of the Louisiana facility were forced to scramble to keep themselves occupied with responsible tasks, and at least a half-generation of inventors looked elsewhere than at coal's problems.

Spectaculars are important not only for a field's self-esteem and attraction of recruits, but also for a reason more subtle. A development engineer is seriously handicapped if he works for years under no great urgency to provide engineering designs for a full-scale plant that he knows will actually be built. There is a coziness in this circumstance hard to resist, which leads to the temptation to resort to dodges that are convenient for getting on with small-scale test work but that are not suited for use in the field. A flaw of some coal development programs in recent years has been too small concern with seeking and testing designs affording coal throughputs per unit volume which are realistic for the commercial scale.

Above all, coal engineers need more coal-engineering establishments — and more coal engineers. There is a strong positive factor in the situation. Anyone associated with the present student generation can tell you that it welcomes the idea of careers with "relevance." Also, engineering students are growing sensitive to career stability. Word of firings from space or military programs reaches campuses quickly, and I predict that engineering enrollments in non-real-world activities will drop precipitately. The space effort has attracted thousands of first-class minds whose loss to the real world's business the nation can ill afford. It will be important to decision makers of the future to obtain good data on the degree to which these minds succeed in obtaining retreads and again finding responsible work. Let us hope the lesson will be learned.

There is a strong element of chance in any youngster's choice of a career, but it is chance directed by events. He sees headlines telling of curiosities and spectaculars. He sees where money is going and what is catching the public fancy. When he reaches college, he will be influenced by the presence of a distinguished professor or a department doing work that is acclaimed. He looks around to see where money is available for fellowships and scholarships and equipment for research.

It has taken a great deal of money to generate the headlines, to produce the fellowships and the fine work at colleges, and to support the symposia and demonstrations, reaching even into secondary schools, which have drawn the first-class people who have carried forward the space and nuclear programs. It will be important to the nation's welfare for coal engineering during the 1970s at last to receive its proper share of these inducements.

Nuclear engineers have a vision whose fulfillment will make plutonium-239 and uranium-233 the "dirty, cheap" fuels and coal the fuel of esteem. It will be valued for the chemicals and clean fuels that can be made from it, and for the metallurgical and electrochemical uses of its fixed carbon. In a coal technology devised to exploit these values, the recovery of sulfur will be a mere inci-

12 Clean Power from Coal, at a Profit

dental. An immediate, properly financed effort to develop means for coping with sulfur can not only give us clean air with profit but also take us a large step along the road toward a coal technology for the twenty-first century.

Postscript

Some readers of the manuscript for this chapter commented that not enough attention was paid to the prospective advantages of the magnetohydrodynamic technique for power generation.

From a thermodynamic standpoint, an MHD generator will stand in exactly the same relationship as a gas turbine to a steam cycle that scavenges heat from hot gases emerging from MHD generator or turbine. Either device for topping the steam cycle can lead to power-generating efficiencies greater than 50%. Adoption of either device can appreciably reduce the amount of heat to be removed from the steam-cycle condenser to the cooling water.

An advantage of the MHD device is that it is less sensitive to presence of dust and volatilized alkali salts in hot combustion gases. The Bureau of Mines[52] is conducting experiments to examine the behavior of K_2O used as a "seed" material to render hot gaseous products of combustion of coal conductive to electricity. The Bureau finds that K_2O is an effective absorbent for SO_2 under the conditions that will exist in an MHD generator. They have demonstrated chemical means for recovering elemental sulfur from K_2SO_4 while regenerating K_2O for reuse.

A disadvantage of MHD is its tendency to produce an undue amount of nitrogen oxides. In a comprehensive design study for the Office of Coal Research (OCR), U.S. Department of the Interior, the Westinghouse Electric Corporation[53] included a large chemical plant to recover nitric acid (as well as sulfuric acid) from MHD generator exhaust gases.

The Bureau of Mines has research in progress which gives hope that nitrogen oxides may be brought under control by employing a two-stage combustion. A first stage, ahead of an MHD generator, would use less than the stoichiometric air for complete combustion, so that no oxygen would be present in gases passing through the generator. Additional air would be added for a secondary combustion, occurring at a much lower temperature, ahead of a waste-heat steam boiler.

In a study for OCR, Avco-Everett Research Laboratory[54] has examined another two-stage combustion scheme for MHD. A first, partial-combustion stage would occur in presence of molten K_2CO_3, which would absorb sulfur to yield K_2S. Ash constituents in the coal would accumulate in the molten K_2CO_3 and would be removed periodically. Sulfur would be recovered from K_2S by a technique regenerating K_2CO_3 for reuse. Oxygen would be employed in both combustion stages to prevent formation of nitrogen oxides, the oxygen being provided from a low-temperature air-fractionation facility.

Note should be taken of several recent developments in SO_2 control technology.

Esso and Babcock & Wilcox have announced a joint effort to develop a regenerable solid absorbent for SO_2 to operate at about 500°C. The Bureau of Mines[55] has published data for a regenerable copper oxide absorbent suitable for use at around 300° to 400°C.

Either of these absorbents appears suitable for use in a panel-bed filter for simultaneous removal of fly ash and sulfur dioxide, under study at The City College.[56,57]

Experiments at The City College[58] have confirmed that half-calcined dolomite is highly reactive toward SO_2 in flue gas at temperatures as low as 600°C. More recent data indicate that reactivity may be adequate even at 500°C, which would allow half-calcined dolomite to be used in a panel-bed filter constructed of carbon steel. Half-calcined dolomite could also serve as an absorbent for SO_2 in a fluidized-bed boiler operated at high pressure and a relatively low temperature, such as the boiler under study by BCURA. The absorbent may be regenerated by a technique yielding elemental sulfur.[58]

NAPCA has announced support of a proposal by Black, Sivalls, and Bryson, Inc.,[59] to study submerged partial combustion of coal in a pool of molten iron. A lime-rich slag resting on the iron would absorb sulfur to form CaS in the slag. Zinc fuming furnaces[60] operate with submerged partial combustion of coal in a slag pool at a superficial velocity of about 5 feet per second. Experience with these furnaces does not suggest that submerged combustion can achieve coal-treating capacities appreciably greater than those attainable in fluidized beds. Since submerged combustion appears less well suited for use at high pressure, it is not as logical a candidate for incorporation in a scheme taking advantage of the cost savings inherent in combinations of gas- and steam-turbine cycles.

References

1
Erik Eriksson, see Chapter 16.

2
C. F. Herington, *Powdered Coal as a Fuel,* 2nd ed. (New York: D. Van Nostrand, 1920), pp. 273–274.

3
R. G. Ramsdell, Jr., and C. F. Soutar, "Anti-pollution program of Consolidated Edison Company of New York," paper presented at ASME Environmental Engineering Conference, Chattanooga, Tennessee, May 1968.

4
K. Wickert, "Experiments on desulfurization before and after the burner for reducing the release of sulfur dioxide," *Mitteilungen der Vereinigung der Grosskesselbesitzer, 83,* 74–82 (April 1963).

5
M. Kruel and H. Jüntgen, "On the reaction of calcined dolomite

and other alkaline earth compounds with the sulfur dioxide of combustion gases as carried out in a cloud of suspended dust," *Chemie-Ingenieur-Technik, 39,* 607–613 (1967).

6

A. V. Slack and H. L. Falkenberry, "Sulfur-dioxide removal from power plant stack gas by limestone injection (plant-scale tests at TVA)," *Combustion, 40,* 15–21 (December 1969).

7

F. Winkler, "Manufacturing fuel gas," U.S. Patent No. 1,687,118 (October 9, 1928).

8

A. A. Godel, "Ten years of experience in the technique of burning coal in a fluidized bed," *Revue générale de thermique, 5,* 349–359 (1966).

9

P. Cosar, "Combustion of coal in a fluidized bed," *Arts et manufactures, 196,* 13–16 (April 1969).

10

H. R. Hoy and A. G. Roberts, "Power generation via combined gas/steam cycles and fluid-bed combustion of coal," *Gas and Oil Power* (July–August 1969), pp. 173–176.

11

J. W. Bishop, E. B. Robison, S. Ehrlich, A. K. Jain, and P. M. Chen, "Status of the direct contact heat transferring fluidized-bed boiler," ASME Paper 68-WA/FU-4 (December 1968).

12

First International Conference on Fluidized Bed Combustion, *Proceedings,* sponsored by the National Air Pollution Control Administration at Hueston Woods State Park, Oxford, Ohio, November 18–22, 1968; a condensed report of the *Proceedings* is available by R. P. Hangebrauck and D. B. Henschel, "Fluid bed combustion, its potential and its R & D requirements: A summary of the First International Conference on Fluidized Bed Combustion," *BCURA Industrial Laboratories Monthly Bulletin, 33,* No. 5, 106–109 (May 1969).

13
J. Highley, ed., *Combustion of Coal in Fluidised Beds,* proceedings of symposium held at Coal Research Establishment, Stoke Orchard, Cheltenham, England (National Coal Board: May 1968).

14
P. L. Waters and A. Watts, "The application of fluidization to coal combustion," paper presented at Canberra meeting of Institute of Fuel (Australian Membership), November 1968.

15
A. R. Cox, L. B. Henson, and C. W. Johnson, "Operation of the San Angelo Power Station combined steam and gas turbine cycle," *Proc. Am. Power Conf., 29,* 401–409 (1967); R. W. Foster-Pegg, "Operation of the San Angelo Power Station combined steam and gas turbine cycle: Discussion," *ibid.,* 410–411.

16
I. I. Kirillov, V. A. Zysin, S. Ya. Osherov, L. V. Aren'ev, and Yu. E. Petrov, "Selection of optimal parameters for the TsKTI-LPI high-temperature gas-steam plant," *Teploenergetika, 14,* No. 1, 44–47 (1967).

17
L. Jéquier, L. Longchambon, and G. van de Putte, "The gasification of coal fines," *J. Inst. Fuel, 33,* 584–591 (1960).

18
A. M. Squires, "Steam-oxygen gasification of fine sizes of coal in a fluidised bed at elevated pressure," *Trans. Inst. Chem. Engrs. (London), 39,* 3–27 (1961).

19
Anonymous, "This new partial oxidation process for making synthesis gas is joining . . . older processes . . . as coal bids for a chemical comeback," *Chemical Week, 78,* No. 26, pp. 76, 78, and 80 (June 30, 1956).

20
L. Reh, "Calcination of aluminum hydroxide in a circulating fluid bed," *Chemie-Ingenieur-Technik, 42,* 447–451 (1970); see also L. Reh and H. J. Ernst, "Possibilities for the calcination of aluminum trihydrate in the fluid bed," *Tagungsheft ICSOBA* (Budapest, October 1969).

21
L. Reh, "Highly expanded fluid beds and their applications," presented at AIChE meeting in San Juan, Puerto Rico, May 1970.

22
A. Rozinek, "Development of the Szikla-Rozinek pulverised fuel gas producer," *Feuerungstechnik, 30,* 153–161 (1942).

23
R. H. Essenhigh and J. M. Beér, "Note: Use of the Szikla-Rozinek combustor as a total gasifier," *J. Inst. Fuel, 33,* 206–207 (1960).

24
W. M. Goldberger, "Collection of fly ash in a self-agglomerating fluidized-bed coal burner," ASME Paper 67-WA/FU-3 (November 1967).

25
P. F. H. Rudolph, "New fossil-fueled power plant process based on Lurgi pressure gasification of coal," *American Chemical Society Division of Fuel Chemistry Preprints, 14,* No. 2, 13–38 (May 1970).

26
C. W. Zielke, H. E. Lebowitz, R. T. Struck, and E. Gorin, "Sulfur removal during combustion of solid fuels in a fluidized bed of dolomite," *American Chemical Society Division of Fuel Chemistry Preprints, 13,* No. 4, 13–29 (September 1969).

27
G. Moss, "The desulfurising combustion of fuel oil in fluidised beds of lime particles," paper at First International Conference on Fluidized Bed Combustion (see reference 12).

28

A. M. Squires, R. A. Graff, and M. Pell, "Desulfurization of fuels with calcined dolomite: I. Introduction and first kinetic results," presented at the AIChE meeting in Washington, D.C., November 1969; to be published in AIChE Symposium Series.

29

W. Guntermann, F. Fischer, and H. Kraus, "Process for removal of sulfur compounds from hot cracked gases," German Patent No. 1,184,895 (January 7, 1965).

30

G. P. Curran, C. E. Fink, and E. Gorin, "CO_2 acceptor gasification process: Studies of acceptor properties," *Advan. Chem. Ser., 69,* 141–165 (1967).

31

M. E. Sacks, C. A. Gray, and R. T. Eddinger, "COED char desulfurization studies," *American Chemical Society Division of Fuel Chemistry Preprints, 13,* No. 4, 287–299 (September 1969).

32

A. M. Squires, "Cyclic use of calcined dolomite to desulfurize fuels undergoing gasification," *Advan. Chem. Ser., 69,* 205–229 (1967).

33

J. D. Batchelor, G. P. Curran, and E. Gorin, "Process for maintaining high level of activity for supported manganese oxide acceptors for hydrogen sulfide," U.S. Patent No. 2,927,063 (March 1, 1960); "Method for maintaining high level of activity . . . ," U.S. Patent No. 2,950,229 (August 23, 1960); "Process for maintaining high level of activity . . . ," U.S. Patent No. 2,950,230 (August 23, 1960); "Manganese ore acceptors for hydrogen sulfide," U.S. Patent No. 2,950,231 (August 23, 1960); "Process for regenerating manganese oxide acceptors for hydrogen sulfide," U.S. Patent No. 3,101,303 (August 20, 1963).

34

F. G. Schultz and J. S. Berber, "Hydrogen sulfide removal from hot producer gas with sintered absorbents," *American Chemical Society Division of Fuel Chemistry Preprints, 13,* No. 4, 30–38 (September 1969).

35

W. G. Behnke, Jr., see Chapter 3.

36

F. W. Theodore, "Low sulfur boiler fuel using the Consol CO_2 acceptor process: A feasibility study," report from Consolidation Coal Company to U.S. Office of Coal Research (November 1967).

37

J. F. Jones, M. R. Schmid, M. E. Sacks, Y. Chen, C. A. Gray, and R. T. Eddinger, "Char oil energy development," report from FMC Corporation to U.S. Office of Coal Research (January 1967).

38

C. L. Tsaros, "Integration of coal-based pipeline gas and power production," *American Chemical Society Division of Fuel Chemistry Preprints, 12,* No. 3, 95–103 (September 1968).

39

R. T. Eddinger, J. F. Jones, and F. E. Blanc, "Development of the COED process," *Chem. Eng. Process, 64,* No. 10, 33–38 (October 1968).

40

C. L. Tsaros, J. L. Arora, and W. W. Bodle, "Sulfur recovery in the manufacture of pipeline gas," *American Chemical Society Division of Fuel Chemistry Preprints, 13,* No. 4, 252–269 (September 1969).

41

N. P. Cochran, "Near term application for MHD power plants," *NEREM Record, 11,* 134–135 (1969); "Production of high quality crude oil and low cost power from Western coals," presented at Las Vegas, Nevada, meeting of Society of Mining Engineers of AIME, September 1967.

42

E. Gorin, G. P. Curran, and J. D. Batchelor, "Desulfurization of carbonaceous solid fuels," U.S. Patent No. 2,824,047 (February 18, 1958).

43

J. D. Batchelor, E. Gorin, and C. W. Zielke, "Desulfurizing low temperature char," *Industrial Engineering Chemistry, 52,* 161–168 (1960).

44

C. A. Gray, M. E. Sacks, and R. T. Eddinger, "Hydrodesulfurization of coal chars," *American Chemical Society Division of Fuel Chemistry Preprints, 13,* No. 4, 270–286 (September 1969).

45

M. Benedict, "Nuclear power prospects," *NEREM Record, 11,* 130 (1969).

46

C. H. Secord, "Slagging grate furnace and method of operation thereof," U.S. Patent No. 3,253,906 (May 31, 1966).

47

H. R. Hoy, A. G. Roberts, and D. M. Wilkins, "Behavior of mineral matter in slagging gasification processes," *Inst. Gas Engrs. J., 5,* 444–469 (1965).

48

B. von Portatius, "The recent development of the Winkler gasification process," *Freiberger Forschungshefte, A69,* 5–25 (1957).

49

W. Peters and H. Bertling, "Kinetics of the rapid degasification of coals," *American Chemical Society Division of Fuel Chemistry Preprints, 8,* No. 3, 77–88 (August 1964).

50

R. Rammler, "Industrial experience with the Lurgi-Ruhrgas process for devolatilization of coal fines," *Erdoel Kohle, Erdgas, und Petrochemie, 19,* 117–121 (1966).

51

M. Gomez, W. S. Landers, and E. O. Wagner, *Entrained-Bed Carbonization of Highly Fluid Bituminous Coals,* Bureau of Mines

Reports of Investigations 7141 (Washington: U.S. Department of the Interior, 1968).

52
D. Bienstock (U.S. Bureau of Mines, Pittsburgh Energy Research Center, 4800 Forbes Avenue, Pittsburgh, Pennsylvania, 15213), personal communication, July 1970.

53
D. Q. Hoover, E. V. Somers, T. C. Tsu, S. Way, W. E. Young, R. W. Foster-Pegg, J. R. Koupal, and H. C. Spohn, "Feasibility study of coal burning MHD generation," report from Westinghouse Electric Corporation to U.S. Office of Coal Research (February 1966).

54
Avco-Everett Research Laboratory, "Study of MHD power system burning char with oxygen," report to U.S. Office of Coal Research (July 1970).

55
D. H. McCrea, A. J. Forney, and J. G. Myers, "Recovery of sulfur from flue gases using a copper oxide absorbent," paper presented at meeting of Air Pollution Control Association, St. Louis, Missouri, June 1970.

56
A. M. Squires and R. Pfeffer, "Panel bed filters for simultaneous removal of fly ash and sulfur dioxide: I. Introduction," *Journal of Air Pollution Control Association,* 20, No. 8, 534–538 (August 1970).

57
L. Paretsky, L. Theodore, A. M. Squires, and R. Pfeffer, "Panel bed filters for simultaneous removal of fly ash and sulfur dioxide: II. Filtration of dilute aerosols by sand beds," paper presented at meeting of Air Pollution Control Association, St. Louis, Missouri, June 1970.

58
A. M. Squires and R. A. Graff, "Panel bed filters for simultaneous

removal of fly ash and sulfur dioxide: III. Reaction of sulfur dioxide with half-calcined dolomite," paper presented at meeting of Air Pollution Control Association, St. Louis, Missouri, June 1970.

59
Anonymous, "Burn coal, save SO_2," *Chem. and Eng. News, 48,* No. 29, 14–15 (July 13, 1970).

60
R. E. Mast and G. H. Kent, "Slag fuming furnaces recover zinc and lead from copper slag," *J. Metals, 7,* 877–884 (1955).

13

**Dealing with Sulfur
in Residual Fuel Oil**

Seymour B. Alpert,
Ronald H. Wolk,
and Arthur M. Squires

In recent years, oil-fired power stations have accounted for about one-tenth of the emissions of SO_2 from power generation in the United States. Many such stations, however, are in urban settings and came under pressure early to reduce SO_2 emissions. Today, as a result of this pressure, a number of power companies in New York, Maryland, Virginia, and Illinois burn oil containing less than 1.0% sulfur. Companies in New Jersey must now burn oil of less than 0.5% sulfur, and companies in Los Angeles must use such oil during seven months of the year. After October 1971, New Jersey's limit will drop to 0.3% sulfur, and New York's Rule 200 will in effect require less than 0.37% sulfur in new power stations. Consolidated Edison Company of New York (Con Edison) plans to convert to 0.37% sulfur oil as quickly as such oil becomes available.

Recent legislative activity might suggest that the target in urban areas will settle down at about 0.3% to 0.5% sulfur in heavy fuel oil. Two standards may evolve: below 0.5% in population centers and moderately higher levels, perhaps 1.0% to 1.5% in less populated regions.[1]

In 1966, power companies burned 23% of the residual oil consumed in the United States, while 27% was used for heating and another 23% by industry. Regulations calling for lower sulfur levels will apply to much of the oil used for heating and industry, as well as to oil used for power. Most users of oil for heating and industry purchase oil in such small amounts that the only practicable way for these users to meet the regulations will be to obtain oil of low-sulfur level from fuel suppliers. A large power company, on the other hand, buys oil in tremendous quantities,

Editors' note. This chapter was received by the editors on July 17, 1970. The references for Chapter 13 are on pages 242–245.

and this fact suggests that the economic answer to this company's SO_2 problem may lie elsewhere. One purpose of this chapter is to argue that this is probably so.

Problems Faced by Fuel Suppliers

Some 627 million barrels of heavy fuel oil were consumed in the United States in 1966 (1 U.S. petroleum barrel = 42 U.S. gallons = 5.146 cubic feet = 158.98 liters). Its combustion contributed about one-third of the SO_2 emissions arising from fossil fuels in that year.

The largest market for fuel oil is on the East Coast. About one-half of the oil burned in the United States is consumed in five states: New York, New Jersey, Pennsylvania, Massachusetts, and Florida.

Imports furnish about 60% of the fuel oil burned domestically, and imports have nearly doubled their share of the market in the past decade, as domestic refiners reduced their yields of heavy oil. In 1966, Venezuelan sources supplied about 90% of the heavy oil burned on the East Coast.

After the 1966 Thanksgiving Day episode in New York City, Con Edison came under intense pressure to reduce its sulfur emissions, and within a year this company lined up supplies of both fuel oil and coal containing not more than 1.0% sulfur. These supplies were of natural origin. Other demands for low-sulfur fuel quickly arose, and these demands too were met by shifting fuels naturally low in sulfur into the critical areas.

Fuel suppliers cannot indefinitely meet the growing demand for low-sulfur fuel by calling upon the world's limited reserves of such fuel. Segregating high-sulfur fuel for nonsensitive markets, such as vessels at sea, can help to release low-sulfur materials for metropolitan consumers, but this approach is limited in usefulness because continuous availability of low-sulfur fuel in large amounts is uncertain. Intense exploration for petroleum now under way in countries such as Libya and Algeria, where low-sulfur oil is already known, cannot be expected to turn up supplies in sufficient amounts to meet the needs of the United States market, let alone those of the world.

Equipment to process high-sulfur fuels will soon be urgently needed, especially if a requirement for fuel of 0.3% to 0.5% sulfur should become general in metropolitan settings. Coal of such low sulfur content is practically nonexistent in the eastern United States, and the requirement would cause a major shift from coal to oil, increasing the demand for low-sulfur oil.

The situation is aggravated by the problem of maintaining imports of Venezuelan residual fuel oil into the United States. This oil contains about 2.1% sulfur and is particularly difficult to convert into a low-sulfur fuel. Revenues derived from the United States market represent a large portion of income for the South American nations that produce and refine this oil, and loss of this market would be disastrous to their economies.

Although investments of several hundred millions of dollars will be required, oil companies give signs that the investments will be made as quickly as technological development makes them prudent.

The technology of desulfurizing residual fuel is in its infancy, perhaps at about the same stage as the catalytic-cracking art of 1940. Trouble has been experienced in several plants that have been built recently, and refiners will wish to see a record of smooth, reliable operation before they make a massive commitment to desulfurization equipment.

We have no doubt that this commitment will be forthcoming. Refiners are just now embarking upon a change in refinery practice which will yield gasoline to meet Detroit's new specification prohibiting use of lead additives. Industrial studies have given estimates ranging from 4 to 11 billion dollars as the investment necessary for changes to eliminate lead. The refiners' willingness to contemplate such huge investments is testimony to their sensitivity to pressures that arise from the public's concern for the environment. It may well be that refiners are anxious to avoid the development of adverse opinion which might put them in jeopardy of legislation disturbing their tax allowance for depletion of oil reserves.

Equipment for catalytic desulfurization of fuel oil is the only means now in sight for providing low-sulfur fuel to the small customer, but we believe other techniques should prove more advantageous in meeting the needs of power generation. In order to understand why this is so, we must first review the available technology for catalytic desulfurization.

Removing Sulfur at the Refinery

Table 13.1 presents inspections of typical virgin heavy fuel oils in general use in the United States, Europe, and Japan. Heavy fuel from Venezuela, a primary source for the American market, typically contains 2.0% to 2.5% sulfur, a lower sulfur content than most Middle East stocks, which are supplied mainly to Europe and Japan. Although Venezuelan oil is lower in sulfur, its catalytic treatment with hydrogen for removal of sulfur is more difficult, since Venezuelan stocks contain higher amounts of the metals

Table 13.1
Inspections of Typical Virgin Fuel Oils

	Venezuelan		Middle East		
Gravity, °API[a]	16.7	11.5	16.1	15.8	22.0
Sulfur, weight percent	2.1	2.8	3.9	3.0	0.4
Pour point, °F	+20	+40	+55	+55	+110
Viscosity, SSF[b] at 122°F	170	275	175	165	65
Vanadium, ppm[c]	250	425	60	36	8
Nickel, ppm	35	70	14	11	1
Volume percent distillate boiling to 975°F	54	49	53	49	55

[a] °API refers to a special, arbitrary hydrometer scale for specific gravity of the American Petroleum Institute. It is widely used and accepted.
[b] SSF is an abbreviation for seconds Saybolt Furol. Viscosity is measured as the time of efflux in seconds of a standard quantity of oil in a Saybolt Furol viscometer.
[c] Parts per million.

vanadium and nickel. When the oil is processed, these metallic constituents poison the catalyst by forming deposits on the catalyst surface. In order to process an oil containing metals, the catalyst must be selected and designed to avoid severe metallic poisoning effects.

An "indirect" technique for catalytic desulfurization of oil has arisen in Caribbean refinery practice and is well suited to supply modest amounts of low-sulfur fuel to the United States market, so long as legislation does not create demand for much fuel below 1.0% sulfur. The indirect technique is attractive because it bypasses the problem of catalyst poisoning by metals. Vanadium and nickel are present in porphyrin compounds that are concentrated in the asphaltene portion of the fuel. In the indirect approach, shown diagrammatically in Figure 13.1, the bottoms of a crude topping unit are distilled under vacuum to remove a heavy gas oil distillate fraction, which is free of vanadium and nickel. This fraction is hydrodesulfurized catalytically to a sulfur content of about 0.2%. The treated fraction is blended back with residual untreated material. Finally, the blend is further cut with low-sulfur fuel from another source, such as low-sulfur North African crude.

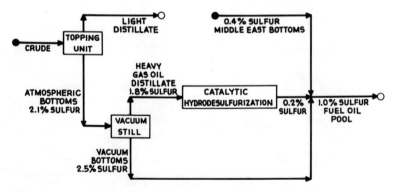

13.1
Schematic diagram of the "indirect" technique for catalytic hydrodesulfurization of residual fuel oil, as it might be applied to Venezuelan atmospheric bottoms to produce oil containing 1.0% sulfur.

The indirect technique is well suited to reduce the sulfur level from 2.0% to 2.5% sulfur to about 1.0% to 1.25% in the overall heavy fuel oil pool.

Several refining companies have received approval from the United States Department of the Interior to import low-cost Venezuelan fuel oil in an amount in excess of their normal quotas, with the understanding that the companies will "comingle" this extra oil with their own atmospheric bottoms and that they will produce from the comingled oil a sweet fuel oil in an amount equivalent to the imported Venezuelan fuel. The point is that Venezuelan residue can be processed at least partially in existing refinery equipment to gasoline and other products of higher value, while domestic bottoms can be processed to supply a low-sulfur residual oil. This approach can provide low-sulfur oil at lower cost than an approach that must deal exclusively with the Venezuelan residue. The economics of the comingling scheme depend on the products that can be obtained and on their sales value. In some situations the refiner can reduce the premium a customer must pay for low-sulfur oil.[2]

The indirect technique for catalytic hydrodesulfurization can be expanded by removing asphalt from vacuum bottoms and desulfurizing the deasphalted material. After the treatment, the high-sulfur asphalt would be blended back with desulfurized oil.

Another possibility would be to coke the vacuum bottoms, but Bechtel Corporation does not consider this economically attractive.[3]

The indirect technique is currently advantageous for oil refiners because plant investment can be stretched out over a period of time. The length of time will depend upon how quickly and how effectively antipollution laws are enforced and upon sulfur levels set by various jurisdictions. Uncertainty in the times at which fuel of lower sulfur level must be supplied, as well as uncertainty in the ultimate target sulfur level, could lead to delays in a refiner's commitment to desulfurization plants. The willingness of fuel buyers and users of power to pay higher costs may also prove a factor in setting the pace of refinery investments for fuel desulfurization.

As demand for low-sulfur oil increases and as legislated sulfur levels become lower, refiners will find themselves compelled to adopt a "direct" approach to catalytic hydrodesulfurization, namely, treatment of substantially the entire atmospheric bottoms in one step.

Processes to carry out direct desulfurization of heavy fuel oil have been offered for license by Hydrocarbon Research, Inc., and Cities Service: H-Oil;[4] Universal Oil Products: RCD Isomax;[5] Chevron: RDS Isomax;[6] Gulf Oil Company: HDS;[7] Esso Research and Engineering and Union Oil Company: RESIDfining;[8] and the Institut Français du Pétrole.[9] There are rumors of other processes under development.

The H-Oil process features an ebullated bed of catalyst in which catalyst particles are buoyed by a rising current of hydrogen and oil undergoing treatment. Other techniques use a number of fixed beds of catalyst. The H-Oil process appears best suited to handle feeds of such high metal content as Venezuelan stocks, although no plant is yet projected for such a feed.

Table 13.2 lists plants for direct desulfurization now in operation or under construction. Most of the plants are for Middle East stocks, and their operation will be followed closely by refiners around the world.

Catalyst technology for direct desulfurization is in an early developmental stage, and improvements can be expected which will reduce the cost of desulfurizing high-sulfur fuel drastically in the next decade.

Removing Sulfur at the Power Station
In the states that consume the greater part of imported Venezuelan oil, power-generating sites of more than 1,000-MW capacity are common. By 1980, new individual power units will commonly

Table 13.2
Plants for Direct Desulfurization of Heavy Residues, in Operation or under Construction

Company	Licensee	Location	Capacity, barrels[a] per day
IdimetsuKosan	UOP[b]	Japan	40,000
Signal Oil	UOP	U.S.	20,000
Aminoil	UOP	Kuwait	35,000
Mizushima	Gulf	Japan	24,000
Kashima	UOP	Japan	40,000
Cities Service	HRI[c]/Cities Service	U.S.	2,500
Kuwait National Petroleum	HRI/Cities Service	Kuwait	28,000
Humble Oil	HRI/Cities Service	U.S.	16,500
Petroleos Mexicanos	HRI/Cities Service	Mexico	18,500

[a] 1 U.S. petroleum barrel = 42 U.S. gallons = 5.146 cubic feet = 158.98 liters.
[b] UOP Universal Oil Products.
[c] HRI Hydrocarbon Research, Inc.

be of a size on the order of 1,000 to 2,000 MW. At these capacities, throughputs of oil are comparable to throughputs in petroleum refining equipment, namely, 37,000 to 74,000 barrels per day for power generation at 100% load factor.

In light of these large throughputs, the question arises, would it be cheaper to remove sulfur in equipment ancillary to power generation than to remove sulfur at the refinery?

To study this question, we must first examine the effect upon the refiner's costs if the power industry should relieve him of part of his burden. In recent years, about one-quarter of the oil burned in the United States has gone to power, and the proportion is probably increasing. What will be the effect upon the cost of three barrels of low-sulfur fuel purchased by a small user if the power industry should gear itself to accept a fourth barrel of high-sulfur fuel?

Figure 13.2 gives a rough answer to this question. The upper curve in the figure is the cost for direct hydrodesulfurization of

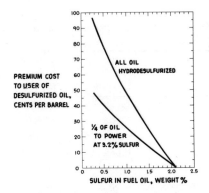

13.2
Costs to provide desulfurized oil, starting from Venezuelan crude. The upper curve is cost for direct hydrodesulfurization of the entire Venezuelan atmospheric bottoms. The lower curve shows the reduction in cost of oil to the small user if the power industry gears itself to accept high-sulfur refinery residues. Costs are for a plant size of 50,000 barrels per day.

Venezuelan atmospheric bottoms to produce fuels of various sulfur levels. For a fuel of 1.0% sulfur, the purchaser would pay a premium of about 50¢ per barrel; for a fuel of 0.3% sulfur, the premium would be nearly $1.00 per barrel. The lower curve in Figure 13.2, on the other hand, shows the premium to be paid by the purchaser of three out of each four barrels of oil if the power industry could accept a fourth barrel at a sulfur content of 3.2%. The power industry's barrel would be a portion left over from a deasphalting operation or from deep vacuum distillation, namely, a distillation carried to 700°C. This barrel would contain substantially all of the metals in the original oil, and the figure shows that its removal would reduce the cost of hydrodesulfurizing the rest of the oil almost by one-half. Obviously, the economic data in Figure 13.2 should represent an interesting, cost-saving alternative for the pollution-conscious public.

For the sake of simplicity, the lower curve in Figure 13.2 has been calculated with the assumption that the power industry would pay the posted price for residual oil of 2.1% sulfur content.

13 Dealing with Sulfur in Residual Fuel Oil

If an enterprising power company executive can line up a supply of high-sulfur oil at a discount appropriate to bunkering fuel, about 75¢ per barrel, the refiner might subsequently find it difficult to collect the full posted price for his high-sulfur discards. Under this circumstance, the economic picture represented in Figure 13.2 would change, but a saving to the public, in price of either electricity or fuel, should remain.

The economic data in Figure 13.2 can be used to argue for any procedure to deal with sulfur at the power station rather than at the refinery. Until recently, attention has been focused upon techniques that remove sulfur at the end of the processing sequence, namely, techniques for removing SO_2 from stack gas. At present, attention is beginning to be paid to the advantages of dealing with sulfur at an earlier stage in the combustion process.

A major opportunity embraces the processing of fuel at high pressure in presence of a desulfurizing agent and the generation of power by a combination of gas- and steam-turbine cycles. An agent derived from limestone or dolomite may advantageously capture sulfur during a first oil-processing step. For this step, there are three cases to consider:

1.
Complete combustion, using air in excess of stoichiometric.
2.
Partial combustion, using between about one-third and one-half of the stoichiometric air and yielding a fuel gas containing CO and H_2.
3.
A cracking operation, yielding low-sulfur liquid products and perhaps also low-sulfur coke as well as fuel gas.[10] Heat for the cracking would be supplied by a partial combustion that consumes about 10% to 15% of the stoichiometric air.

If complete combustion is chosen, there will be an advantage in selecting a temperature and pressure of combustion such that $CaCO_3$ will not decompose. For example, a combustion at 800°C and a pressure greater than about 6 atmospheres would lead to a CO_2 partial pressure sufficient to preserve $CaCO_3$. Although limestone would not be reactive toward SO_2 at these conditions, half-calcined dolomite would react readily:

$$[CaCO_3 + MgO] + SO_2 + 0.5O_2 = [CaSO_4 + MgO] + CO_2 \quad [1]$$

Kinetics for this reaction, apparently because of half-calcined dolomite's porosity, are favorable at temperatures even as low as 600°C.[11-13]

The combustion could advantageously be conducted in a fluidized bed of half-calcined stone. About 70% of the heat release from the combustion would be removed by heat transfer to boiling water in steaming tubes that are in contact with the fluidized bed. The remaining heat would appear as sensible heat in combustion gases leaving the bed and passing to a gas turbine.

Work at The City College has shown that $[CaCO_3 + MgO]$ can be regenerated from $[CaSO_4 + MgO]$ by a technique that liberates H_2S for sulfur manufacture.[13] The sulfate would be reduced to $[CaS + MgO]$ by action of H_2, and sulfur would be desorbed from CaS by reacting the solid with steam and CO_2 at high pressure:[14]

$$[CaS + MgO] + H_2O + CO_2 = [CaCO_3 + MgO] + H_2S \quad [2]$$

A partial combustion would possess at least two advantages. A fluidized bed of either fully or half-calcined dolomite for partial combustion would be about one-half as large as a bed for complete combustion, because the quantity of gas leaving the bed would be less by about one-half. Since gas from the partial combustion would contain fuel values in the form of H_2 and CO, a secondary combustion outside the bed could provide hotter gas to the gas turbine. As gas-turbine machinery becomes available for use at higher temperatures, it will be advantageous to convert as much of the fuel's energy as possible into sensible heat in combustion gases at the inlet to the gas turbine and as little as possible into steam raised in the first fuel-treating step.

13 Dealing with Sulfur in Residual Fuel Oil

Reaction 2 could be used to recover sulfur from [CaS + MgO] formed in a fluidized bed of calcined dolomite for partial combustion.

Cracking has major advantages over a partial combustion that converts all of the fuel values into a fuel gas containing simply H_2 and CO. The gas yield would be smaller if cracking were used, and the volume of high-pressure oil-treating equipment would be correspondingly less. Cracking could provide low-sulfur fuel products that can be stored, so that the oil-treating equipment could operate steadily at a fixed oil throughput which is smaller than the throughput needed to operate a customer power station at full rating.

Figure 13.3 illustrates broadly a scheme under study at The City College[10] for processing residual fuel oil in equipment ancillary to a customer power station. Partial combustion, to provide heat of cracking, may consume either oil or coke, a product of cracking. Cracked fuel vapors and petroleum coke may be desulfurized by

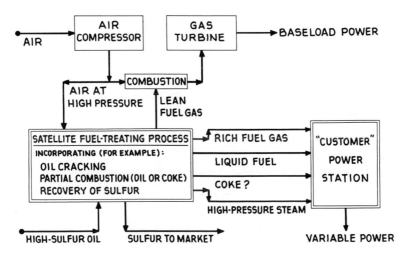

13.3
Schematic diagram of equipment satellite to a power station for supplying the station with low-sulfur fuels.[10]

cooperative action of H_2 and calcined dolomite, and sulfur may be recovered by means of Reaction 2. Hydrogen would arise autogenously within the process. Lean fuel gas from the partial combustion may be used to drive a gas turbine. Nearly one-half of the fuel products could be liquid or solid, which may be stored against variable demands for power, while the fuel-treating equipment would operate steadily at a fixed throughput of oil. For a customer station having an overall average load factor of 70%, the station could be turned down to 37% of its rating before the clean fuel plant need be cut back. The scheme is nearly a break-even proposition from an energy standpoint, and its adoption might well lead to a lower cost for electricity.

The capital cost for direct catalytic hydrodesulfurization equipment to handle the entire Venezuelan atmospheric bottoms and to provide oil of 0.3% sulfur will be on the order of $500 per daily barrel of throughput. This works out to be an investment of about $13 per kilowatt for a power station operating at a 70% load factor and burning the desulfurized oil. Estimates indicate a considerably lower capital cost for the scheme depicted in Figure 13.3, where no hydrogen plant is needed, processing pressures are far lower, and the operation is noncatalytic. In addition, gas-turbine power-generating equipment has a capital cost well below that of steam power equipment, and base-load power generated by the gas turbine in Figure 13.3 will be a bargain.

Dutch Shell[15] and several groups in Russia[16-18] have studied the partial combustion of oil with air at elevated pressure to provide fuel gas for combinations of gas- and steam-turbine cycles.

Although far less work has been done upon ideas such as these than upon ideas for removing SO_2 from stack gas, many engineers are beginning to recognize that the former ideas hold greater promise for controlling SO_2 emissions at reasonable cost.

So many enthusiastic announcements have appeared concerning so many varied schemes for removing SO_2 from stack gas that even persons closely following the technology have sometimes acquired the impression that a number of the schemes are now ready for commercial use. On the contrary, a distinguished panel of engineers, organized under the auspices of the National Academy of Engineering, recently concluded that *"contrary to widely held belief, commerically proven technology for control of sulfur oxides from combustion process does not exist."* [19] The panel judged that a process must pass the test of operation on the scale of 100 MW or larger before it could be judged commercially proved.

Two schemes have reached field trials at this scale. The Tennessee Valley Authority is conducting trials of a dry limestone-injection technique. Early results have been disappointing, sulfur removals generally running less than 30% even with application of several times the stoichiometric amount of stone for absorption of the total sulfur. Combustion Engineering is testing a wet limestone system that uses substantially the stoichiometric amount of stone. First trials ran into difficulties at the Meramec Station of Union Electric. A different design has given better results in preliminary operation at the Lawrence Station of Kansas Power & Light Company.[20]

Commonwealth and Boston Edison companies have recently announced trials on a scale larger than 100 MW for schemes in which stack gas will be scrubbed with a water slurry of an alkaline reagent, lime and magnesia, respectively. Several other schemes are perhaps also ready for such trials.

An overblown heralding of these schemes before completion of the trials could lead to pressure for their application on an imprudently wide scale. The history of classic disasters of engineering — for example, the *Great Eastern,* the postwar Fischer-Tropsch synthesis, the Fermi reactor —should teach caution in applying new technology on a giant scale. Too much money and hope committed to multiple installations of inadequately tested systems could make it difficult to fund work on more advanced schemes for SO_2 control. The National Academy panel[19] warned: "Efforts to force the broad-scale installation of unproven processes would be unwise; the operating risks are too great to justify such action, and there is a real danger that such efforts would, in the end, delay effective SO_2 emission control." The panel concluded: *"A high level of government support is needed for several years to encourage research, engineering development, and demonstration of a variety of the more promising processes."*

The panel felt that public money would not be needed to develop processes for desulfurizing fuel oil. Indeed, the oil industry can be expected to work vigorously upon the problem of developing oil-hydrodesulfurization techniques capable of handling the total atmospheric bottoms of Venezuelan crude. We cannot feel certain, however, that the oil industry will view it to be in its interest to develop a technology for dealing with refinery high-sulfur discards in equipment ancillary to power generation, permitting the economic savings depicted in Figure 13.2 to be realized. Government or power industry funds will probably be needed to support such developments.

References

1
C. W. Siegmund, "Low sulfur fuels are different," *Hydrocarbon Processing, 49,* No. 2, 89–95 (February 1970).

2
A. R. Johnson, L. Lehman, and S. B. Alpert, "Petrochemical feedstocks from residuals," paper presented at 158th National ACS Meeting, New York City, September 7–12, 1969.

3
Bechtel Corporation, "Desulfurization costs: Residual fuel oil," report for American Petroleum Institute, February 1967.

4
Anonymous, "Hydrocracking: The H-Oil process," *Petrochemical Engineer, 41,* No. 5, 39–52 (May 1969).

5
M. Yamato and C. Watkins, "Commercial RCD Isomax processing of residual oils," *Proceedings, American Petroleum Institute Division of Refining, 48,* 80–84 (1968).
6
J. Scott, A. Bridge, and R. Christensen, "The Chevron RDS Isomax process," paper presented at Japan Petroleum Institute Fuel Oil Desulfurization Symposium, Tokyo, March 10–17, 1970.
7
A. Henke, "Gulf's residual HDS process," Japan Petroleum Institute Fuel Oil Desulfurization Symposium, Tokyo, March 10–17, 1970.
8
J. Blume, D. Miller, and L. Nicolai, *Hydrocarbon Processing, 48,* No. 9, 131–136 (September 1969).
9
F. Audibert and P. Duhaut, "Residual reduction by hydrotreatment," paper presented at 35th Midyear Meeting of American Petroleum Institute Division of Refining, Houston, Texas, May 15, 1970.
10
A. M. Squires, "System to provide clean power from residual oil," paper presented at meeting of Air Pollution Control Association, New York City, June 22–26, 1969.
11
J. R. Coke, "The removal of sulfur oxides from waste gases by a dry method," unpubl. diss. (mentor: M. W. Thring), University of Sheffield, England, Department of Fuel Technology (May 1960).
12
R. R. Bertrand, A. C. Frost, and A. Skopp, "Fluid bed studies of the limestone based flue gas desulfurization process," Interim Report, Contract No. PH 86-67-130, for National Air Pollution

Control Administration, Esso Research and Engineering Co., Linden, New Jersey (October 31, 1968).

13
A. M. Squires and R. A. Graff, "Panel bed filters for simultaneous removal of fly ash and sulfur dioxide: III. Reaction of sulfur dioxide with half-calcined dolomite," paper presented at meeting of Air Pollution Control Association, St. Louis, Missouri, June 14–18, 1970.

14
A. M. Squires, "Cyclic use of calcined dolomite to desulfurize fuels undergoing gasification," *Advan. Chem. Ser., 69,* 205–229 (1967).

15
T. K. De Haas, J. K. Nieuwenhuizen, M. Akbar, J. A. van der Giessen, and L. W. ter Haar, "Prevention of air pollution by sulphur dioxide," paper presented at Tokyo Section Meeting, World Power Conference, October 1966.

16
V. S. Altshuler and G. V. Klirikov, "Gasification of heavy, high-sulfur-content oil at pressures up to 70 bar to obtain power gas," *Teploenergetika, 11,* No. 4, 70–73 (1964).

17
A. I. Andryushchenko, P. S. Chernyshev, V. L. Polishchuk, V. A. Ponyatov, A. I. Popov, and L. A. Slyusarev, "Effectiveness of steam-gas plant burning high-sulfur oil," *Teploenergetika, 13,* No. 1, 16–20 (1966).

18
M. I. Derbaremdiker, M. K. Pismen, P. M. Sharov, K. L. Serebrennikova, and N. I. Nikolaev, "Gasification of oil under pressure in an air blast," *Teploenergetika, 13,* No. 6, 22–25 (1966).

19
Ad Hoc Panel on Control of Sulfur Dioxide from Stationary Combustion Sources, "Abatement of sulfur oxide emissions from

stationary combustion sources," Report COPAC-2, Committee on Air Quality Management, Committees on Pollution Abatement and Control, Division of Engineering, National Research Council, Washington, D.C. (1970).

20
Anonymous, "Modified SO_2 system faces new tests," *Electrical World* (December 8, 1969), pp. 38–39.

14

Climatic Consequences of Increased Carbon Dioxide in the Atmosphere

Gordon J. F. MacDonald

Man, through the burning of fossil fuels, has significantly increased the carbon dioxide content of the atmosphere. Further, carbon dioxide plays a key role in determining the thermal balance of the atmosphere. Climate, in turn, is fixed by the thermal balance existing within the atmosphere. The complexities of global climate are still too poorly understood to state with any degree of confidence the effect of carbon dioxide produced by human activity on climatic perturbations. However, both historic and geologic evidence show that significant variations have taken place in the past in the global climatic regime. Since a highly industrialized society will continue to require energy, derived in substantial part from the fossil fuels, it is essential that we understand the effects of continually loading the atmosphere with carbon dioxide.

In these pages I will first review evidence with regard to the changing content of atmospheric CO_2. Then I will examine the sources of CO_2 and attempt to forecast what changes in atmospheric composition we may expect in the future. The effects of CO_2 on the thermal balance of the atmosphere will be compared with the observed climatic changes of the past sixty years. In addition, since the earth's climate is influenced by the yearly production of man-made energy, we need to be concerned not only with CO_2 resulting from industrialization but with the overall thermal budget of the atmosphere. These considerations lead to a number of important policy questions.

Observed Changes in the Atmospheric Content of Carbon Dioxide

The burning of fossil fuels (primarily lignite, coal, petroleum,

The references for Chapter 14 are on pages 261–262.

14 Climatic Consequences of Increased CO_2 in the Atmosphere 247

and natural gas) now releases about 1.5×10^{16} grams of CO_2 in one year. This figure should be compared with the amount consumed annually in photosynthesis of about 1.1×10^{17} grams. Thus the combustion of fossil fuels today produces an amount that is an appreciable fraction (approximately 1/7) of the CO_2 entering the photosynthesis cycle each year. (See Figure 14.1.) It is, furthermore, about four orders of magnitude greater than the return of carbon to the fossil reservoir; the rate of release of CO_2 into the atmosphere by oxidation of recently grown organic materials matches to within one part in 10^4 the amount consumed in photosynthesis.

Until recently, it was not clear how much of the CO_2 being released by combustion accumulated in the atmosphere and how much entered the oceans and the terrestrial biomass. Callendar calculated that CO_2 had increased at approximately a constant rate from the nineteenth-century level of about 290 parts per million (ppm) to about 330 ppm in 1960.[1] In order to sustain such an increase, about three-fourths of the amount released through combustion would have to remain in the atmosphere. More recently, Keeling has undertaken a detailed monitoring program at Mauna Loa Observatory in Hawaii and at the Pole station in Antarctica. He has found that the annual average CO_2 levels measured at Mauna Loa and at the Pole station agree to within 1 ppm. Furthermore, both stations show a consistent increase over the past few years. Keeling finds a concentration of CO_2 of about 314 ppm and that the rate of increase averages about 0.2 part per million per year (ppm/yr).[2] This implies that each year the mass of CO_2 in the atmosphere increases by 5×10^{15} grams. Of all the CO_2 produced by combustion, one-third of it remains in the atmosphere and two-thirds are taken up by the oceans and by the biomass. At the current rate of deposition in the atmosphere, the amount of man-made CO_2 doubles every 23 years.

CO_2 is one of the three important radiation-absorbing con-

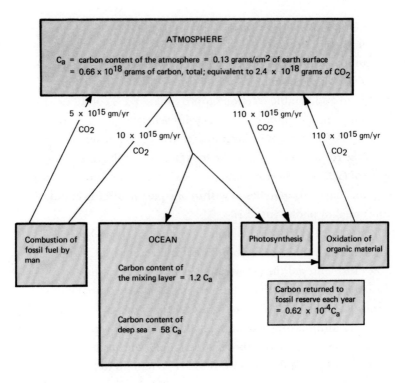

14.1
Carbon resources of the earth.

stituents of the atmosphere. The other two are water vapor and ozone. While man-made activities may have affected the latter two constituents, the changes in their concentration are much smaller than the change in CO_2. If man is changing the thermal balance of the atmosphere through the introduction of gaseous pollutants, it is through the addition of carbon dioxide.

Sources of Carbon Dioxide

At present, the world's use of energy is increasing annually by about 4%, which corresponds to a doubling time of 17 years. Of all the energy produced, about 98% comes from oil, coal,

14 Climatic Consequences of Increased CO_2 in the Atmosphere   249

and natural gas, while water power contributes only 2%; on a global scale nuclear energy is still negligible. At present, the total thermal energy produced from crude oil, coal, and lignite equals 3.8×10^{12} kilowatt hours per year (kWh/yr), which corresponds to 8.5×10^{-3} watts per square meter (W/m²) averaged over the surface of the earth. This can be compared with the energy production in 1940 of 3.1×10^{-3} W/m² averaged over the earth's surface.

Not only has the total amount of energy used by modern industrial society increased very greatly, but also the sources of the energy have changed. Since 1800, the principal sources of the world's industrial energy have been the fossil fuels and waterpower. Before 1900 the energy derived from oil, as compared with that obtained from coal, was almost negligible. Since 1900 the contribution of oil to the total energy supply has steadily increased. It is now approximately equal to that of coal and is increasing more rapidly. If natural gas and natural gas liquids are added to crude oil, the petroleum group of fuels represents about 60% of all the energy derived from coal, petroleum, and waterpower. Since World War II the production of coal has grown at a constant rate of about 3.6% with a doubling period of 20 years. World production of crude oil, except for a slight retardation during the depression of the 1930s and during World War II, has increased from 1890 to the present at a nearly constant exponential rate of 6.9% per year with a doubling period of 10 years.

The rates of growth of a highly industrialized nation differ substantially from the world average. The U.S. production of total energy from coal, oil, natural gas, and waterpower divides into two distinct growth periods. From 1850 to 1907 energy in the United States was produced at a growth rate of 7% per year with a doubling period of 10 years. From 1907 to the present, the growth rate dropped to 1.8% per year with a doubling period of 39 years.

Keeling's observations on the current increase of CO_2 in the atmosphere, together with the fact that about one-third of the CO_2 released through combustion enters the atmosphere, implies that man's activities have added about 1.7×10^{17} grams of carbon dioxide to the atmosphere. In considerations of climate, it is not only the total amount of CO_2 introduced into the atmosphere that is important but also the rate at which it is introduced. Although coal has been mined for about 800 years, one-half of the coal produced during that period has been mined during the last 31 years. Half of the world's cumulative production of petroleum has occurred during the 13-year period since 1956. Thus, 9×10^{16} grams of CO_2 have been introduced since 1950, and 1.3×10^{17} grams have been introduced since the mid-1930s.

An estimate of the maximum amount of CO_2 that man might introduce into the atmosphere can be made from estimates of the total fossil fuels that are available. Averitt estimates that 7.6×10^{12} metric tons of coal represents the maximum available source.[3] This is about twice the coal resources established by actual geologic mapping. Estimates of petroleum resources vary considerably. Weeks and Ryman estimate that approximately 2×10^{12} barrels of oil are ultimately recoverable.[4,5] Hubbert appears to favor the somewhat lower figure of 1.35×10^{12} barrels.[6] (1 U.S. petroleum barrel = 42 U.S. gallons = 158.98 liters.) If these fossil fuels were burned, they would add to the atmosphere 3.3×10^{18} grams of CO_2. This figure should be compared with the 2.2×10^{18} grams now in the atmosphere. Thus man is capable of increasing the current CO_2 content by about 150%.

Hubbert estimates that 80% of the ultimate resources of the petroleum family (crude oil, natural gas, tar, sand oil, and shale oil) will be exhausted in about a century.[6] The time required to exhaust 80% of the world's coal resources would be 300 to 400 years but only 100 to 200 years if coal rather than nuclear power is used as the main energy source. These rates of consumption imply that, in the next century, the content of the CO_2 in the atmosphere could be doubled.

Projected possible increases of carbon dioxide in the atmosphere must be viewed with caution. The oceans, for example, contain 60 times more carbon dioxide than does the atmosphere. If the atmosphere's average temperature increased, either as a result of the increasing carbon dioxide content or through some other means, the heating would tend to drive some of the carbon dioxide now dissolved in the oceans back into the atmosphere. Alternatively, the increasing flow of nutrients into the ocean, resulting from improper agricultural practices, would stimulate the growth of phytoplankton and increase, through photosynthesis, removal of carbon dioxide from the atmosphere. Looking into the future, we cannot say with any precision what the carbon dioxide content of the atmosphere actually will be. What we can say is that man has changed the total amount of carbon dioxide by several percent and, further, that he is capable of more than doubling the carbon dioxide content over the next hundred years.

Thermal Effects of Increasing Carbon Dioxide
A number of attempts have been made to calculate the effect of carbon dioxide on the average surface temperature. The energy, which drives the atmosphere and determines climate, is derived primarily through the absorption of visible solar radiation by the earth's surface and atmosphere. The absorption of that energy tends to raise the temperature of the surface, which maintains its thermal balance by reradiating energy to space at longer wavelengths. The absorption of incoming visible radiation by carbon dioxide is so small that changes in its concentration will have no appreciable effect on the amount of incoming solar radiation that reaches the surface. However, carbon dioxide is opaque to certain parts of the long-wave

radiation emitted by the earth's surface and atmosphere; changes in concentration of carbon dioxide change the quantity of heat lost by radiation into space. The size of the temperature change resulting from the change in concentration of carbon dioxide depends additionally on the water vapor concentration in the atmosphere and its cloudiness.

The most complete calculations of the net effect of altering the carbon dioxide content of the atmosphere are those of Manabe and Wetherald.[7] These numerical calculations show that an increasing content of CO_2 results in the warming of the entire lower atmosphere, the amount of warming being dependent in part on whether the atmosphere is held at fixed absolute humidity or at fixed relative humidity. With the assumption of fixed absolute humidity, together with conditions of albedo, cloudiness, radiation, and other parameters chosen as typical of the mid-latitudes, an increase of 10% in the CO_2 concentration leads to a warming of 0.2°C (degrees centigrade). For the assumption of a fixed relative humidity, with all other conditions remaining the same, a 10% increase in content of CO_2 raises the temperature by 0.3°C. Manabe and Wetherald consider the fixed relative humidity conditions as being more realistic. If the amount of carbon dioxide in the atmosphere is doubled at constant relative humidity, the temperature is increased by 2.4°C.

Observed Climatic Changes

There are three basic sources of data on climatic conditions of the past. The most exact data cover the period of instrumental observations that began on a large scale during the second half of the nineteenth century. The historical observations provide information on changes with a time scale measured in hundreds and thousands of years. Finally, data on climate of more remote eras, extending over hundreds of millions of years, are furnished by paleogeographic studies as well as observations on the distribution of certain isotopes.

14 Climatic Consequences of Increased CO_2 in the Atmosphere

Climate is much too complicated to be described by a single parameter, but a useful guide is the temperature of the atmosphere measured at the earth's surface and averaged over the year and over the whole earth. The last advance of the ice sheet in Eurasia ended about 10,000 years ago. And since then, the permanent ice cover in the Northern Hemisphere has been limited largely to the Arctic Ocean and to some islands in the high latitudes. In the last 10,000 years, thermal conditions in the higher and middle latitudes have continued to change and, over the last century when instrument observations became available, the fluctuations continued.

Meteorological data show that from 1880 to 1940 the average temperature increased by about 0.6°C while in the last 25 years the average temperature has decreased by about 0.2°C. Associated with the increasing temperature were northward movements of the frost and ice boundaries; pronounced aridity in south central parts of Eurasia and North America leading to dust-bowl conditions; and strong northern hemispheric zonal circulation. In more recent times, the lowering temperature is associated with frost and ice boundaries shifting to the south; a weakening zonal circulation; and marked increases in rainfall in parts of previously arid continental areas.

The recent years have not been exceptions. Sea-ice coverage in the North Atlantic in 1968 was the most extensive in over 60 years. As a result, Icelandic fishermen suffered great losses. In contrast, the rains in central continental regions, particularly in India, led to very high wheat yields. These observations further emphasize the point that the complex pattern of human activity is very sensitive to relatively small changes in climate.

Many workers have associated the increasing temperature of the early part of the century with the increasing CO_2 content. Consideration of the numerical effects of increased CO_2 and the period of rapid growth of CO_2 content in the atmosphere suggests that only a small fraction of the changes observed prior to 1940 can be attributed to increased CO_2. Because of the relatively short time scale for doubling of energy consumption, only about one-fourth of the CO_2 that man has used was added to the atmosphere prior to 1940. Since the net effect of increasing the content of CO_2 by 10% is to increase the average temperature by 0.3°C at fixed relative humidity, then, prior to 1940, CO_2 could have increased the temperature only by something less than 0.1°C. On the other hand, during a period in which very substantial amounts of carbon dioxide have been added to the atmosphere, the temperature has been dropping; thus the heating effects of carbon dioxide over the last 30 years have been compensated for either by natural fluctuations or by other man-made activities.

It should be noted that there are at least five ways, in addition to increasing the carbon dioxide concentration, in which man's activities could significantly perturb the atmospheric heat balance and thus the climate:

1.
Decreasing atmospheric transparency by aerosols resulting from industry, automobiles, and home-heating units.
2.
Decreasing atmospheric transparency by dust put into the atmosphere as a result of improper agricultural practices.
3.
Direct heating of the atmosphere by the burning of fossil and nuclear fuels.
4.
Changing the albedo (the percentage of the incoming solar radiation directly reflected outward) on the earth's surface through urbanization, agriculture, and deforestation.
5.
Altering the rate of transfer of thermal energy and momentum between the oceans and atmosphere by oil film resulting from incomplete combustion and oil spills from oceangoing vessels.

The effects of three of these — urban pollution, agricultural pollution, and changing the albedo — tend to lower the temperature, while direct heating of the atmosphere tends to increase the temperature. The long-term consequences of altering the surface of oceans are uncertain. The fact is that we do not know the magnitudes of these effects. Thus we are uncertain as to the extent to which man is, indeed, changing the climate of his planet.

Overall Thermal Budget of the Atmosphere
Emission of CO_2 to the atmosphere affects the thermal budget in an indirect way by holding within the atmosphere some fraction of the long-wave radiation. Industrial activities, however, can affect climate in a much more direct way. The energy used by industry is ultimately turned into heat. At present, this energy is being obtained as a result of photosynthesis that occurred in past epochs; in the future, the energy may be derived in a variety of new ways. What is important in the considerations of climate is that this energy represents a new source of heat in addition to the heat of solar radiation. The heat thus produced today affects the microclimate of large cities. In the future, it may affect the climate of the planet itself.

The energy budget for the earth is illustrated in Figure 14.2. The mean annual difference between absorbed solar radiation and long-wavelength radiation into space is about 68 watts per square meter (W/m^2). Most of this radiation balance is used up in the evaporation of water, heating of the atmosphere, and driving various meterological processes. A tiny part, less than 1%, is used in the photosynthesis of green plants and is turned into a relatively stable form of chemical energy. Averaged over the surface, man at present is producing about 8.4×10^{-3} W/m^2, or somewhat more than one ten-thousandth of the radiation balance of the atmosphere. Although this fraction is much too small to affect climate on a large scale, it certainly alters local

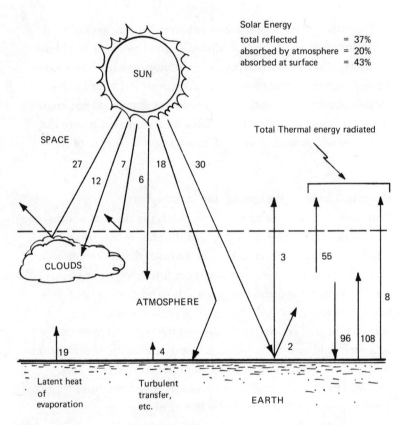

14.2
Approximate energy budget for the earth. The numbers represent percent of 1,400 watts/meter^2, which is approximately the solar constant. The solar energy numbers pertain to a disc earth of area πr_e^2. The remaining energies pertain to a spherical earth of area $4\pi r_e^2$. In the figure, these numbers have been multiplied by 4 to keep all units the same.

microclimate. In the future, direct input of energy into the atmosphere may be of major importance.

In a primitive society, the utilization of energy is principally limited to the food consumed by the individual. This corresponds to about 100 thermal watts per person. The present world average is somewhat in excess of 1,000 thermal watts per person, while a highly industrialized society such as the United States uses about 10,000 thermal watts per person. Projecting into the future, we see that if the world population grows to 5 billion and if the worldwide average of energy use is 10,000 thermal watts per person, then the direct energy input into the atmosphere would be 0.1 W/m², or about one-hundredth that of the natural radiation balance. Indeed, if the present rate of energy production of 4% per year is maintained, then in 200 years artificial energy input into the atmosphere would equal one-third of the radiation balance. This level would be reached in only a hundred years with a 10% increase of energy. As Budyko has argued, only an increase of a few tenths of one percent in the radiation balance would be sufficient to cause the polar ice to disappear entirely.[8] Thus the combined effect of carbon dioxide pollution and direct heat pollution is strongly in the direction of warming the earth.

Conclusions

The problem of carbon dioxide pollution of the atmosphere illustrates several generalizations that hold with respect to the alteration of the environment by man. These include the following:

1.
Large-scale man-made changes may be taking place in the environment. Some of these changes, such as the direct heat pollution of the atmosphere, have been recognized only recently.

2.
The magnitude of the changes produced by man is of the same order as that caused by nature. The carbon dioxide added to the atmosphere can bring about a change of several tenths of a degree in the average temperature, and changes of this magnitude have been observed.

3.
The alterations produced by man can no longer be regarded as

local. Direct heat input by a city changes the microclimate of that city. The combined effects of many cities are changing the climate of the planet.

4.

Our understanding of the physical environment is sufficient to identify inadvertent modification but is far too primitive to predict confidently all the consequences of man's unwise use of his resources.

5.

Inadvertent as well as purposeful modification of our environment is a neglected area of research, neither fashionable to the scientists and engineers nor, until recently, of high priority to the money-distributing government agencies. For example, there are at most a handful of small research groups in the country studying, in a professional way, the influence of man's activities on the climate despite the very great importance of understanding long-term changes in our environment. Of the monies provided by the federal government for research in weather modification, about 1% support programs in inadvertent modification.

I believe that urgent action is required if the nation is to deal responsibly with the long-term problems of climate alteration. Such action includes the following:

1.

There should be worldwide recognition, both at the national levels and within the United Nations, of the long-term significance of man-made alterations on climate.

2.

Worldwide programs should be developed for ground monitoring of atmospheric turbidity, carbon dioxide content, and water vapor distribution, with particular attention to oceanic areas. The ground-based observation should be supplemented by airborne monitoring of the number, density, and size distributions of the particulate matter, and the composition of both the particulate matter and gaseous constituents of the atmosphere. Such programs are an essential first step to establish base levels of pollution against which changes in the decades ahead can be measured.

3.

Satellite programs should be developed continually to monitor, on a global basis, the cloud cover and the heat balance of the atmosphere. There should be a continuing program for measuring surface albedo, with a particular emphasis on changes brought by man-made alteration.

4.

It is urgent that computer simulation of climate should be intensified. Particular attention should be devoted to the effects of altering the thermal balance by changes in the albedo, CO_2, and dust particles.

5.

The federal government should assign high priority to research in problems of inadvertent modification. Such research should examine not only the question of monitoring the geophysical and industrial parameters affecting climate but also the problems of constructing more adequate models of the thermal and the dynamic processes within the atmosphere as well as at the boundaries between the atmosphere, the solid earth, and the oceans.

Only if we undertake a research program commensurate with the problem can we assure ourselves that the many small inputs into the atmosphere from man will not lead either to a disastrous ice age or to an equally disastrous melting of the polar ice caps.

Discussion

Question from David MacLean, Associate Professor of Chemistry, Gustavus Adolphus College, St. Peter, Minnesota.

Has the recent increase in dust content of the air, related to an increase in volcanic activity, caused a cooling of the earth's climate?

Answer by Gordon J. F. MacDonald.

It is not at all clear what have been the relative contributions of volcanic dust and industrially introduced aerosols to the observed cooling of the past three decades. Mitchell has argued that volcanic dust dominates over industrial waste products and that, at least in principle, the observed cooling is due to the increased volcanic material in the stratosphere.[2] Before this question can be resolved with any sense of definiteness, we need much better information on the distribution, as a function of altitude, of particles and their composition, as well as their sizes. These kinds of data can then be introduced into radiative models of the atmosphere to calculate in detail the climatic effect.

Question from Mary C. Roberts, an active conservationist and a member of the Sierra Club.

What are the effects on climate of widespread deforestation, creation of large numbers of small farm ponds, and other changes in the landscape associated with agriculture?

Answer by Gordon J. F. MacDonald.

The effects of deforestation are of two kinds. Deforestation changes the albedo — the percentage of the sun's radiation that is reflected directly back out into space — and also changes the surface properties. A change in the surface properties leads to a different rate of interchange of heat and momentum with the surface. Increasing albedo, which might be expected through extensive deforestation and perhaps subsequent formation of desert or arid regions, will lead to an average decrease in the surface temperature of the earth. We do not know the overall effect of urbanization and improper agricultural practices on the earth's albedo. The advent of satellite technology will help immeasurably in that satellites can maintain a continuous record of the average albedo.

Question from Delvin S. Fanning, Associate Professor of Soil Science, University of Maryland.
What is known of man-induced changes in the oxygen content of the atmosphere?
Answer by Gordon J. F. MacDonald.
At present there is no evidence that man is bringing about measurable changes in the oxygen content of the atmosphere. However, such changes might be expected if continued supply of phosphates, nitrates, and other minerals affect in a substantial way the ecological balance in near-shore oceanic areas and large lakes.

References
1
G. S. Callendar, "On the amount of carbon dioxide in the atmosphere," *Tellus, 10,* 243–248 (1958).
2
C.D. Keeling, as reported by J. Murray Mitchell, Jr., "A preliminary evaluation of atmospheric pollution as a cause of the global temperature fluctuations of the past century," in *Global Effects of Environmental Pollution,* S. Fred Singer, ed. (Dordrecht, Holland: D. Reidel Publishing Co., 1970).
3
Paul Averitt, *Coal Resources of the United States, January 1, 1967,* Geological Survey Bulletin 1275, U.S. Department of the Interior (Washington: U.S. Government Printing Office, 1969), 116 pp.
4
Lewis G. Weeks, "Fuel reserves of the future," *Bulletin of the*

American Association of Petroleum Geologists, 42, No. 2, 431–438 (1958).

5

W. P. Ryman, quoted by Hubbert (see reference 6).

6

Marion King Hubbert, "Energy resources," in *Resources and Man,* published for the Committee on Resources and Man of the National Academy of Sciences — National Research Council (San Francisco: W. H. Freeman and Co., 1969). (This book was originally assigned the NAS Publication No. 1703. This assignment was in error, and the number 1703 has since been assigned to another publication.)

7

Syukuro Manabe and Richard T. Wetherald, "Thermal equilibrium of the atmosphere with a given distribution of relative humidity," *J. Atmospheric Sciences, 24,* 241–259 (1967).

8

M. I. Budyko, "Changes of climate," *Meteorologiya i gidro'ogiya,* No. 11, 18–27 (1967).

15

Atmospheric Chemistry Richard D. Cadle
 and Eric R. Allen

The chemistry of the atmosphere above an altitude of about 80 kilometers is dominated by ions and electrons; that below this altitude is concerned largely with neutral molecules, free radicals, and atoms. The composition and reactions of the former region, the ionosphere, were recently reviewed by Donahue.[1] Therefore, we shall restrict ourselves to a discussion of the photochemistry of the lower part of the atmosphere.

There are many reasons for interest in the chemistry of the lower atmosphere. One of these, of course, is strictly academic: we would like to satisfy our curiosity as to this aspect of the way our universe behaves and the way our atmosphere evolved. Another is that the concentrations of many minor constituents vary considerably from one atmospheric region to another and thus can be used to follow atmospheric motions. Chemical reactions are often responsible for the formation and removal of such constituents (ozone, O_3, for example), and a knowledge of the photochemistry involved is essential to the correct interpretation of concentration measurements. Studies of this sort have provided much information concerning large-scale atmospheric motions, information that may eventually aid in the development of numerical methods for long-range weather forecasting.

A third reason relates to air pollution. A prevalent type of pollution ("photochemical smog") has very unpleasant properties that are largely the result of substances produced by photochemical reactions involving the contaminants. Preventing the emission of all such contaminants into the atmosphere is impractical at present, and an understanding of the chemical reactions that

Editors' Note. This chapter was originally published as an article, "Atmospheric Photochemistry," in *Science, 167,* 243–249 (1970), copyright 1970 by the American Association for the Advancement of Science. It is reproduced with permission.
References for Chapter 15 are on pages 285–288.

produce the unpleasant substances is essential to the effective selection of pollutants on which to concentrate the greatest control efforts. Also important is a knowledge of the ultimate fate of pollutants since this affects pollution on a worldwide basis. This fate is often that the pollutants undergo further photochemical reactions.

The chemical and dynamic behavior of the atmosphere is markedly influenced by its temperature structure, which is often used to establish atmospheric regions lying one above another (Fig. 15.1). The lowest region of the atmosphere is the troposphere, which extends from the earth's surface up through the region of generally decreasing temperature to the tropopause, the boundary between the troposphere and the stratosphere. The tropopause decreases in altitude with increasing latitude but at mid-latitudes is at about 12 kilometers. The troposphere, as we all know, is a region of storms and turbulence and receives most of its heat from the ground rather than directly from the sun. The stratosphere is a region of nearly constant or increasing temperature with increasing altitude, and it extends upward to about 50 kilometers. Above the stratosphere, starting at the stratopause, is the mesosphere in which the temperature falls until an altitude of 80 to 85 kilometers is reached. Farther out is the thermosphere, a region of rising temperature to at least 200 to 300 kilometers; it is often considered to blend with the solar corona. Thus our discussion is restricted to the chemistry of the troposphere, stratosphere, and mesosphere.

Photochemical reactions can be defined as those in which the initiating step is the absorption of a photon by an atom, molecule, free radical, or ion. This absorption can produce excited species, decomposition (photolysis), or ionization. A minimum photon energy (frequency) is required for each process and each species. The initial effect of photon absorption can be considered to be the primary photochemical reaction. Subsequent reactions initiated by primary products are called secondary photochemical

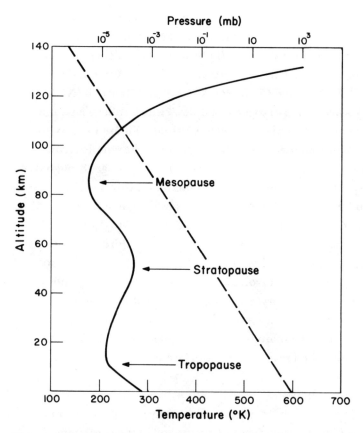

15.1
Schematic diagram of air temperature (solid line) and pressure (broken line) as a function of height.[34]

reactions, and almost all atmospheric reactions involve both types of reaction.

A large assortment of techniques is used to study the chemistry of the atmosphere, but in general these can be classified into three types. One involves the qualitative and quantitative determination of the composition of atmospheric samples. Near the ground, fairly conventional chemical and instrumental analytical techniques can usually be applied, although special methods are sometimes required where the concentrations are very low, for example, in the range of parts of constituent per billion parts of air by volume (ppb). Samples at higher altitudes are collected from balloons, aircraft, or rockets. A recent development at the National Center for Atmospheric Research (NCAR) is the rocket-borne "cryogenic sampler," which collects a sample of air between 40 and 60 kilometers altitude (Figure 15.2). With liquid hydrogen used as the coolant, most of the sample is condensed as it is rammed into the device during passage through the atmosphere. The second class of techniques involves indirect methods of analysis. For example, the absorption or scattering of solar ultraviolet radiation by ozone can be used to determine ozone concentrations at various altitudes. The third class consists of laboratory studies of reactions believed to occur in the atmosphere. Some of these involve attempts to reproduce the atmosphere in the laboratory. Such attempts have often been successful, especially in air pollution research. Other laboratory studies have been designed to obtain fundamental quantitative information on the individual reactions that can be applied to the atmosphere.

The Troposphere
The chemistry of the troposphere is mainly that of a large number of minor atmospheric constituents and of their reactions with molecular oxygen. The photochemical behavior is dominated by the fact that essentially no solar radiation of wavelength less than 2900 angstroms reaches the tropopause, radiation of the shorter wavelengths having been removed largely by reaction with ozone in the stratosphere and with molecular oxygen. Above 3300 angstroms, however, the solar spectral intensity distribution is similar to that at the top of the atmosphere for a sun directly overhead. At large zenith angles, that is, with the sun near the horizon, attenuation is observed over all wavelengths as a result of scattering, diffusion, and the increased path length for ozone in the atmosphere. Typical curves for the distribution of solar spec-

15 Atmospheric Chemistry

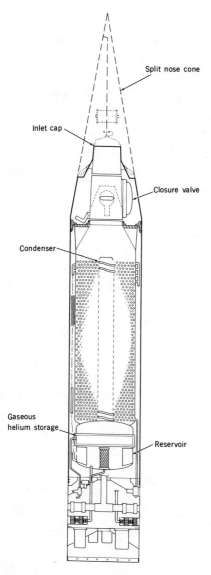

15.2
National Center for Atmospheric Research cryogenic sampler. The cooling agent is liquid hydrogen pressurized with helium. The sample is condensed when air is rammed into the device at an altitude between 40 and 60 kilometers.

tral intensity are shown in Figure 15.3 for $z = 0°$ and $z = 80°$, at ground level in mid-latitudes. Nitrogen does not absorb radiation that reaches the troposphere, and molecular oxygen absorbs only very weakly in the red end of the visible spectrum. Thus the primary photochemical processes in the troposphere are almost entirely restricted to minor constituents.

The atmospheric chemistry of sulfur compounds has been studied extensively because they are unpleasant constituents of smog and pollute the atmosphere worldwide. Also, their variation in concentration has posed some challenging questions with regard to large-scale motions. For example, it is not at all clear why the relative concentrations of airborne sulfate particles are exceptionally high in polar regions.[2] The sources of such compounds include domestic and industrial fuel consumption, volcanoes, forest fires, and bacterial action.[3] The oceans also may be an important source of sulfur dioxide. The most abundant of these substances are hydrogen sulfide, sulfur dioxide, sulfuric acid,

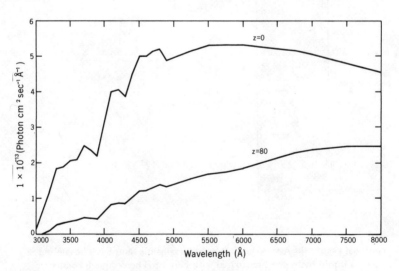

15.3
Estimated solar spectral intensity distribution at ground level and mid-latitudes for zenith angles 0° and 80°.[21,35]

and various sulfates. The concentrations are extremely variable but are as much as several parts per million (ppm) in severe smog.

Hydrogen sulfide and sulfur dioxide undergo a number of oxidation reactions. Ozone is produced in photochemical smog, and some also mixes into the troposphere from the stratosphere, where it is a natural constituent. Hydrogen sulfide reacts with ozone to form mainly sulfur dioxide and water but at rates probably too slow, at least in the gas phase, to be important.[4] Hydrogen sulfide has a bond energy of 81.1 kilocalories per mole per H-S bond, corresponding to a wavelength (λ) of 3,510 angstroms. However, it does not appreciably absorb radiation of wavelength greater than about 2,700 angstroms; thus it does not undergo photolysis or react photochemically with molecular oxygen. Hydrogen sulfide, ozone, and molecular oxygen are all soluble in water. The rate of oxidation of hydrogen sulfide by these oxidants in solution, for example, in fog or cloud droplets, may be very fast, but these reactions have not been studied. Atomic oxygen, produced largely by the photolysis of ozone and of nitrogen dioxide, is present in extremely small amounts in the troposphere. In spite of the low concentrations, atomic oxygen reactions almost certainly occur to an appreciable extent, especially in some polluted atmospheres. The reaction of atomic oxygen with hydrogen sulfide[5] can be represented by the equations

$$H_2S + O \rightarrow OH + HS \qquad [1]$$

and

$$-d[H_2S]/dt = k[H_2S][O],$$

where k, "the rate constant," equals 4×10^{-14} cubic centimeters per molecule per second. The brackets refer to concentrations expressed in numbers of molecules or atoms per cubic centimeter, and times is in seconds.* Reaction 1 is followed by a chain reaction leading to products such as SO_2, SO_3, H_2SO_4, H_2, and H_2O.

---

* Many rate constants are temperature dependent. However, for purposes of comparison, only values for ~300°K are given in this chapter. The rate constants presented were in many cases selected from among several given in

Sulfur dioxide in its electronic ground state does not react with ozone in the gas phase either in the presence or in the absence of water vapor.[6] The energy for the dissociation of SO_2 into SO and O is about 135 kilocalories per mole ($\lambda < 2{,}100$ angstroms). However, SO_2 absorbs strongly in the near-ultraviolet region to form electronically excited SO_2, which may react with ozone. This reaction has not been studied. Similarly, ground state SO_2 does not react with O_2 in the gas phase. However, excited sulfur dioxide does react, and several laboratory studies have been made of this reaction.[7] The rate at which such a reaction occurs in the atmosphere can be estimated from the photochemical yield Φ (the ratio of molecules of product formed per photon absorbed), the absorption coefficient of the primary reactant, the intensity of solar radiation, and the concentration of the reactants. Integration is necessary over the wavelength range in which absorption occurs. Unfortunately, a very wide range of values for Φ has been obtained. Recent results obtained in our laboratory indicate that Φ is about 1.7×10^{-2} at 3,130 angstroms, but over the wavelength range from 2,800 to 4,200 angstroms, Φ is about 2×10^{-3}. This difference may partially explain the previous discrepancies. The latter value corresponds to an atmospheric reaction rate of about one SO_2 molecule per thousand oxidized per hour.

Two nonphotochemical means of oxidation of SO_2 by O_2 are of possible importance. Urone et al.[8] have found that in the presence of powdered oxides of aluminum, calcium, chromium, iron, lead, and vanadium, sulfur dioxide in air is oxidized very rapidly without sunlight. Airborne dust particles may have a similar effect. Sulfur dioxide dissolved in fog or cloud droplets is oxidized by

---

the literature. Although considerable judgment was used in the selection, there was often little basis for choice among several values. If we use the product of the rate constant and reactant concentrations, the rate of disappearance of reactants or formation of products may be estimated as above. The rates obtained in this way are used as a guide to the relative importance of simultaneous reactions occurring in the atmosphere.

O_2 much more rapidly than by the gas-phase photochemical reaction, especially when certain metal salts are present.[9]

Sulfur dioxide undergoes a three-body reaction with atomic oxygen:[10]

$$SO_2 + O + M \rightarrow SO_3 + M \qquad [2]$$

where M is a molecule of O_2 or N_2, and $k = 1.3 \times 10^{-32}$ cm^6 molecule^{-2} sec^{-1}. This reaction is more important in stratospheric and in polluted air than in the "natural" troposphere. The sulfur trioxide formed almost immediately reacts with water vapor to form sulfuric acid, which in turn will rapidly react mainly with any ammonia present to form ammonium sulfate or ammonium bisulfate.[11] Thus sulfur trioxide never exists to an appreciable extent in the open atmosphere.

Nitrogen compounds, such as ammonia, nitrous oxide, nitric oxide, nitrogen dioxide, and nitric acid vapor, are interesting minor constituents of the atmosphere. Several of these undergo oxidation and hydrolysis, the end products being nitric acid and nitrates. Ammonia is liberated from the soil and the sea and is probably removed from air, where it exists both as the gas and as ammonium salts, mainly as a result of the scrubbing action of rainfall. Except for reaction with acids, ammonia is chemically quite stable in the natural troposphere. The reaction with ozone has not been studied but is probably very slow. Ammonia does not absorb visible or ultraviolet radiation of wavelengths exceeding 2,200 angstroms and thus does not undergo excitation or photolysis below the ionosphere. It does react with atomic oxygen:[12]

$$NH_3 + O \rightarrow NH_2 + OH \qquad [3]$$

where $k = 1.7 \times 10^{-15}$ cm^3 molecule^{-1} sec^{-1}.

Nitrous oxide (N_2O) is the prevalent oxide of nitrogen in the atmosphere, occurring throughout the lower atmosphere at about 0.3 to 0.5 ppm. The main source of nitrous oxide seems to be soil bacteria, since reactions of ground state or excited O, O_2, or O_3 with N_2 cannot account for it.[10,13] In the lower atmosphere N_2O is chemically quite inert.

Nitric oxide (NO) is the main oxide of nitrogen produced by combustion of organic material in air (by "nitrogen fixation"), and combustion is probably the chief source below the ionosphere. Its concentration in the troposphere away from major sources is probably of the order of 10^{-2} ppm. It is very slowly oxidized by O_2 but rapidly by O_3:[14]

$$O_3 + NO \rightarrow NO_2 + O_2 \qquad [4]$$

where $k = 5 \times 10^{-14}$ cm^3 molecule^{-1} sec^{-1}. Nitric oxide does not absorb solar radiation that reaches the troposphere. Atmospheric NO_2 is produced mainly by the oxidation of NO. Little is known concerning its concentration in unpolluted air, although Lodge and Pate[15] found comparable concentrations of NO and NO_2 in the tropics.

When nitrogen dioxide is irradiated with near-ultraviolet radiation, O_2 and NO are formed:[16]

$$NO_2 + h\nu \rightarrow NO + O \qquad [5]$$
$$NO_2 + O + M \rightarrow NO + O_2 + M \qquad [6]$$

where $k = 4.2 \times 10^{-31}$ cm^6 molecule^{-2} sec^{-1}, and $h\nu$ represents a photon of light. However, in air containing less than 1 ppm of NO_2 and 21% O_2, the following reaction will be faster than Reaction 6:[17]

$$O + O_2 + M \rightarrow O_3 + M \qquad [7]$$

where $k = 7.5 \times 10^{-34}$ cm^6 molecule^{-2} sec^{-1}. Since Reaction 7 is followed by Reaction 4, steady-state concentrations of NO_2 and O_3 may be maintained at times, although these are usually disturbed by transport of ozone from the stratosphere and from photochemical smog formed as discussed in the next section.

Ozone also reacts with NO_2, although not so rapidly as with

NO. The reaction occurs in two steps:

$$NO_2 + O_3 \rightarrow NO_3 + O_2 \qquad [8]$$
$$NO_3 + NO_2 + M \rightarrow N_2O_5 + M \qquad [9]$$

The rate is controlled by the slower reaction, Reaction 8. The rate constant for this reaction was found by Ford, Doyle, and Endow[18] to be 3.3×10^{-17} cm^3 molecule^{-1} sec^{-1}. The N_2O_5 reacts very rapidly with water vapor to form nitric acid vapor, HNO_3.

Nitrogen dioxide hydrolyzes in the gas phase:

$$3NO_2 + H_2O \rightleftarrows 2HNO_3 + NO \qquad [10]$$

This reaction has been investigated by McHaney,[19] who found the following equilibrium constant:

$$K = \frac{[HNO_3]^2[NO]}{[NO_2]^3[H_2O]} = 0.004/\text{atm} \qquad [11]$$

at 300°K. The equilibrium lies far to the left, but in air, with its high concentration of water vapor, about 5% of the NO_2 is converted to HNO_3 at room temperature.

Various organic compounds are present in the natural troposphere, the most abundant being methane. Concentrations of methane, butane, acetone, n-butanol, and carbon monoxide were found to be 1.6 ppm, 0.06 ppb, 1.0 ppb, 190 ppb, and 90 ppb, respectively, in uncontaminated arctic air masses at Point Barrow, Alaska.[20] Many plants evolve organic vapors such as hydrocarbons, esters, aldehydes, and ketones.

Methane in the atmosphere is of considerable interest chemically. It is evolved by the natural and polluted environment. Natural sources include anaerobic decay of vegetation in the biosphere and natural gas seepage. Atmospheric concentrations away from sources of pollution average about 1.5 ppm. Methane does not absorb solar radiation above a wavelength of 1,600 angstroms and is chemically quite inert in the troposphere. Considerable evidence exists that some of the other organic compounds undergo photochemical reactions similar to those occurring in smog which are described in the next section. Such reactions may be responsible for the formation of the haze often observed over forests.

The Polluted Atmosphere

It is difficult to divorce a discussion of chemical reactions occurring in a localized polluted environment from tropospheric chemistry as a whole. The difference is mainly a matter of concentration. Unfortunately, it takes exposure to severe local pollution and observation of the accompanying unpleasant effects, such as eye irritation, odor, plant damage, and reduced visibility, to make us acutely aware of the profuse contamination of the atmosphere. We should, however, also be aware that, when conditions are not conducive to the maintenance of a stagnant atmospheric layer, the same contaminants are pumped at the same rate into the troposphere but, fortunately, are diluted by diffusion and mixing into a considerably larger volume.

Although several types and combinations of pollution are observed, we shall confine our discussion to pollution of the type first recognized in the Los Angeles basin and often called photochemical smog. We now know that the unpleasant properties of the smog over practically all cities are caused at least in part by compounds produced by photochemical reactions.

It is well established that the main contaminants in photochemical smog originate from automobile exhaust, although certainly there is a contribution from industrial activity and the incineration of wastes (Table 15.1). The exhaust gases consist primarily of nitrogen oxides, nitrogen, uncombusted and partially combusted hydrocarbons from the fuel, oxides of carbon, and water vapor. Carbon oxides and water vapor are relatively inert to solar radiation at the ground and therefore are not involved in the primary photochemical processes, but they may play a role

Table 15.1
Typical concentrations of trace constituents in photochemical smog; parts of constituent per hundred million parts of air by volume (pphm)

Constituent	Concentration (pphm)
Oxides of nitrogen	20
NH_3	2
H_2	50
H_2O	2×10^6
CO	4×10^3
CO_2	4×10^4
O_3	50
CH_4	250
Higher paraffins	25
C_2H_4	50
Higher olefins	25
C_2H_2	25
C_6H_6	10
Aldehydes	60
SO_2	20

in secondary reactions.* For a detailed review of the data available prior to 1961, the reader is referred to the excellent treatise by Leighton.[21] The limitations on the size of this article preclude more than a general description of the major processes involved. Obviously, the interactions are numerous and complex, and the problem is compounded by the large variety of reactants.

Variation in the concentration of major reactive species with time of day during a typical smoggy day in Los Angeles is shown in Figure 15.4. The sequence of formation, destruction, and dispersion is easily observed. The rate of emission of pollutants into the atmosphere varies throughout the day and peaks around midmorning and late afternoon. The solar radiation intensity also varies continuously, increasing in the morning and decreasing in the afternoon.

* Of course, CO_2 is involved in plant photosynthesis.

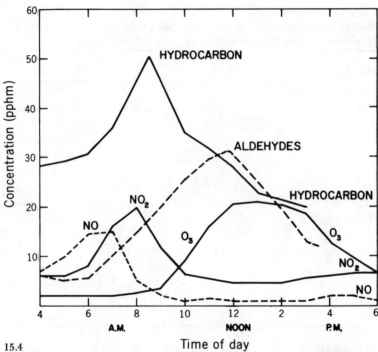

15.4 Time of day
Typical variation of components in photochemical smog for a day of intense smog;[21] pphm, parts of constituent per hundred million parts of air by volume.

During the daylight hours, the nitric oxide is rapidly converted to nitrogen dioxide. The mechanism of this rapid conversion and that of the production of ozone (Figure 15.4) are two of the major problems in understanding the development of smog. Oxidation of the nitric oxide directly by O_2 is much too slow a reaction to account for the conversion, and, as pointed out earlier, nitric oxide does not absorb solar radiation in the troposphere.

Laboratory experiments suggest that the most important primary photochemical reaction in smog may be the photolysis of nitrogen dioxide (Reaction 5), which is always formed to a slight extent by the direct reaction with O_2. Most of the oxygen atoms produced (>99%) react with molecular oxygen (Reaction 7) to produce ozone, but this is followed by the very rapid Reaction 4, which leads to the regeneration of NO_2. The result, if these were the only reactions occurring, would be a steady-state concentration of ozone much lower than that observed in smog.[22] However, a small part of the atomic oxygen reacts with the hydrocarbons and other organic compounds to produce a wide variety of organic and inorganic free radicals, for example:

$$O + \text{olefins} \rightarrow R + R'O \text{ or } \overset{O}{\overset{\diagup \diagdown}{R - R'}} \qquad [12]$$

where $k = 1.7$ to 8.0×10^{-12} cm^3 molecule^{-1} sec^{-1}, and R and R' are organic radicals. The ozone also reacts with various organic compounds, especially olefins, to form a multitude of compounds, including free radicals:

$$O_3 + \text{olefins} \rightarrow \text{products} \qquad [13]$$

where $k = 3$ to 30×10^{-18} cm^3 molecule^{-1} sec^{-1}.

The organic free radicals react rapidly with O_2 to form peroxy free radicals:[21]

$$R + O_2 \rightarrow RO_2 \qquad [14]$$

where k is approximately equal to 10^{-14} cm^3 molecule^{-1} sec^{-1}. These are capable of oxidizing NO to NO_2 and may be responsible for most of the conversion of NO to NO_2 in smog.

Aldehydes are products of the reaction of both ozone and atomic oxygen with olefins. They can react with ozone and atomic oxygen, which initiates new chain reactions. Generally, ozone reacts with aldehydes to produce peroxy acids. Atomic oxygen, on the other hand, abstracts hydrogen to produce a hydroxyl and an acyl radical.

The reaction of oxygen atoms with acetaldehyde has been studied in this laboratory:[23]

$$O + CH_3CHO \rightarrow CH_3CO + OH \qquad [15]$$

where $k = 4.5 \times 10^{-13}$ cm^3 molecule^{-1} sec^{-1}. Both of these products are capable of reacting further and maintaining a reaction chain. For example, the hydroxyl radical may react with the relatively inert carbon monoxide[24] by way of the reaction

$$\text{OH} + \text{CO} \rightarrow \text{CO}_2 + \text{H} \qquad [16]$$

where $k = 1.7 \times 10^{-13}$ cm^3 molecule^{-1} sec^{-1}. The hydrogen atoms thus produced will react almost exclusively with oxygen to form hydroperoxy radicals[25] by way of the reaction

$$\text{H} + \text{O}_2 + \text{M} \rightarrow \text{HO}_2 + \text{M} \qquad [17]$$

where $k = 3 \times 10^{-32}$ cm^6 molecule^{-2} sec^{-1}, and the hydroperoxy radicals are capable of oxidizing nitric oxide to nitrogen dioxide. Also, the acyl radicals will probably react with oxygen to produce acylperoxy radicals that are capable of oxidizing nitric oxide.

Finally, it has been suggested that excited molecular oxygen may be produced by energy transfer from photoexcited hydrocarbons.[26] This oxygen may then react with nitric oxide, olefins, or aldehydes to produce nitrogen dioxide directly or indirectly.

The ozone in smog may be produced largely by the reaction of peroxy radicals with O_2:[22]

$$\text{RO}_2 + \text{O}_2 \rightarrow \text{RO} + \text{O}_3 \qquad [18]$$

Many other components absorb solar radiation at the wavelengths we have considered, for example, nitric acid, nitrates, nitrites, nitro compounds, aldehydes, ketones, peroxides, acyl nitrates, and particles. All that can be said at this time is that the absorption of solar radiation by these compounds will contribute to the production of free radicals, which in turn will produce new compounds.

The products that are particularly noxious to plant and animal life are reasonably well understood. Formaldehyde, peroxyacyl nitrate, and acrolein are probably the main components responsible for eye irritation. Formaldehyde is a product of the reaction of ozone or atomic oxygen with terminal olefins; so, for that matter, is acrolein. Peroxyacyl nitrate, on the other hand, is the termination product of a reaction chain and requires the addition of nitrogen dioxide to a peroxyacyl radical.

The photooxidation of sulfur dioxide occurs by way of processes mentioned in the section dealing with the chemistry of the natural troposphere.

The Stratosphere and the Mesosphere
The photochemistry of the stratosphere and the mesophere differs considerably from that of the troposphere largely because of the presence of shorter-wavelength, more energetic, solar radiation (down to about 2,000 angstroms at the top of the mesophere, although there is a small amount below 2,000 angstroms, for example, the so-called Lyman α band at 1,216 angstroms).

The chemical reactions in the stratosphere and mesosphere are largely the result of the photolysis of O_2 to form atomic oxygen. This photolysis is produced by absorption in the strong Schumann-Runge bands in the wavelength range from 1,760 to 2,030 angstroms and the weaker absorption extending to wavelengths of about 2,450 angstroms:

$$O_2 + h\nu \rightarrow O + O \qquad [19]$$

The following reaction sequence then occurs, leading to the formation and destruction of ozone:[27]

$$O + O + M \rightarrow O_2 + M \qquad [20]$$

where $k = 2.8 \times 10^{-33}$ cm^6 molecule^{-2} sec^{-1}; atomic oxygen reacts with molecular oxygen to form ozone (Reaction 7);

$$O + O_3 \rightarrow 2O_2 \qquad [21]$$

where $k = 1.9 \times 10^{-15}$ cm^3 molecule^{-1} sec^{-1}; and

$$O_3 + h\nu \, (\lambda < 11,400 \text{ Å}) \rightarrow O_2 + O \qquad [22]$$

Reactions 7 and 21 constitute the principal mode of recombination of O atoms in this region. Numerous investigations have been made of the rate constants and photochemical yields of these reactions. These and the absorption coefficients for ozone and oxygen have formed the basis for numerous calculations of concentrations of atomic oxygen and of ozone as a function of altitude in the stratosphere and mesosphere. Perhaps the greatest uncertainties in these calculations, other than atmospheric transport effects, are the activation energy for Reaction 21 and the extent to which the absorption of utraviolet radiation by O_2 and O_3 obeys the Beer-Lambert law.

Calculations should probably also include reactions involving hydrogen compounds and free radicals. Unfortunately, the chemistry then becomes extremely complicated. Photochemical dissociation of water vapor in the mesosphere, oxidation of methane and other hydrocarbons in the stratosphere, and reactions of electronically excited O from ozone photolysis in the ultraviolet are eventual sources of hydrogen atoms which may react by at least 30 processes; the two most important processes in the stratosphere are Reactions 17 and 23:[25,28]

$$H + O_3 \rightarrow OH + O_2 \qquad [23]$$

where $k = 2.6 \times 10^{-11}$ cm^3 molecule^{-1} sec^{-1}. The process represented by Reaction 23 has been suggested as being responsible for the airglow emission by the hydroxyl radical. Two other processes of particular importance are:[29]

$$OH + O \rightarrow H + O_2 \qquad [24]$$

where $k = 5 \times 10^{-11}$ cm^3 molecule^{-1} sec^{-1}, and

$$HO_2 + O \rightarrow OH + O_2 \qquad [25]$$

where k is greater than or equal to 10^{-11} cm^3 molecule^{-1} sec^{-1}. These reactions are of some importance in the mesophere where atomic oxygen concentrations are not less than 10^{10} per cubic centimeter during the day. Generally, in the mesosphere the atomic oxygen concentration exceeds that of ozone during the day; the reverse is true in the stratosphere.

A considerable number of secondary processes, incorporating the destruction of atomic hydrogen, hydroxyl, and perhydroxyl, occurs in the mesosphere. Incorporation of all possible reactions into a complete scheme to describe the mesosphere would be a monumental task and could only be attempted by the use of a high-speed computer. The complexity of this region is demonstrated by Figure 15.5, which indicates the concentrations of some of the constituents as a function of altitude.

The photolysis of ozone produces electronically excited O_2, at least for $\lambda < 2,500$ angstroms. The role that such oxygen plays in the chemistry of the stratosphere and mesosphere is unknown.

Numerous comparisons have been made of the calculated and measured concentrations of ozone. The differences have provided much useful information concerning atmospheric motion.

Little direct information exists concerning nitrogen compounds (other than N_2) in the stratosphere and mesosphere. Ammonium compounds have been detected in the particulate material collected from the lower stratosphere, for example, by Cadle et al.[30] These were probably formed by the reaction of ammonia with droplets of sulfuric acid produced as described in the next paragraph. Nitrous oxide is chemically stable throughout this region. Nitric oxide, which enters the stratosphere from below, is rapidly oxidized by ozone to NO_2, which in turn is oxidized to N_2O_5. The end product of these reactions (Reactions 4, 8, 9, and the reaction of N_2O_5 with water vapor) is nitric acid vapor, which has been detected spectroscopically in the stratosphere.

Small amounts of H_2S and SO_2 must reach the stratosphere from the troposphere and be oxidized, largely by reaction with atomic oxygen (Reactions 1 and 2). However, the only sulfur compounds that have been detected in this region are particulate sulfates and persulfates. Some of these particles are produced in

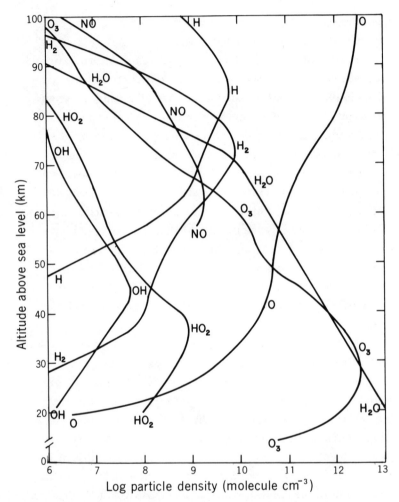

15.5
Daytime equilibrium profiles for a hydrogen-oxygen atmosphere. A composite of experimental data from the *Handbook of Geophysics* (New York: Macmillan, 1961) and the theoretical estimates of Hunt.[36] The nitric oxide profile was measured by Pearce.[37] The great number of reactions and secondary processes that are possible make this region of the atmosphere extremely complex.

the troposphere or at the earth's surface, but most are probably formed in the stratosphere by chemical reaction, such as Reactions 1 and 2 followed by reaction of the SO_3 with water vapor and ammonia to form ammonium sulfate.[3,10] Cadle et al.[30] showed that the concentration of sulfate immediately above the tropopause is considerably greater than that just below it. A very interesting feature of the stratosphere is the existence of a worldwide layer containing an especially large concentration of such particles at an altitude of about 18 kilometers.[3] The concentration seems to be markedly increased after major volcanic eruptions, presumably as a result of the emission of large amounts of sulfur compounds into the atmosphere.

The main carbon compounds in the stratosphere and mesosphere are carbon monoxide, carbon dioxide, and methane. The chief source of carbon monoxide is the combustion of organic material, especially by the internal-combustion engine. Carbon monoxide is relatively inert in the troposphere, but in the stratosphere it is removed by reaction with O and OH, especially the latter (Reaction 16).[24] Although numerous studies have been made of the reaction of O with CO, there is considerable uncertainty concerning the kinetics of this reaction.

Carbon dioxide is both created and destroyed at ground level, the main source being the combustion of fossil fuels. Scavenging is accomplished mainly by the action of planets, bacteriological action, and solution in the sea. Carbon dioxide is transported vertically by mixing processes and presumably is destroyed above the mesopause by photochemical decomposition, which requires radiation of shorter wavelength than 1,650 angstroms.

Present estimates of CO_2 concentrations near ground level average about 330 ppm. Although there is some evidence that it decreases by about 5 ppm above the tropopause, recent measurements made by Scholz et al.[31] in the vicinity of the stratopause indicate a CO_2 concentration of about 320 ppm. This value was obtained for a sample of air collected with the cryogenic sampler between 45 and 62 kilometers. Concern has been expressed over several decades about the CO_2 abundance, which apparently is increasing at the rate of 0.7 to 1 ppm per year. The strong influence of CO_2 on the stratospheric radiation budget suggests that increased heating will cause major climatological changes to occur at the earth's surface. Fortunately or unfortunately, this seems to have been offset or even reversed by a cooling effect due to the accumulation of particulate matter in the atmosphere.

A limited number of vertical profiles of methane have been obtained by Bainbridge and Heidt.[32] They found a nearly constant mixing ratio (the concentration relative to O_2 and N_2) below the tropopause that decreased at and above the tropopause. These results were extended by Kyle et al.[33] Scholz et al.[31] estimated that there was less than 0.1 ppm of CH_4 in a sample of air collected with the cryogenic sampler. Cadle and Powers[10] suggested that the primary process responsible for removal of the methane was reaction with stratospheric atomic oxygen:

$$O + CH_4 \rightarrow CH_3 + OH \qquad [26]$$

where $k = 5.5 \times 10^{-15}$ cm^3 molecule^{-1} sec^{-1}. In this way, all hydrocarbons that reach stratospheric levels would be eventually oxidized to CO_2 and water. Furthermore, this reaction may be an important source of OH.

Summary
Photochemical reactions in the troposphere, stratosphere, and mesosphere are, to a large extent, reactions among minor constituents of the atmosphere. The chemistry is markedly limited by the minimum wavelength of the solar radiation that penetrates to a given atmospheric level. It is useful to differentiate between the chemistry of city smog and that of the ambient atmosphere, but the entire atmosphere is being polluted and the difference is one of degree.

Both laboratory and field studies are contributing to our knowledge of atmospheric photochemistry, and the most convincing conclusions have been obtained by combining the results of the two methods of investigation.

References

1
T. M. Donahue, *Science, 159,* 489 (1968).

2
R. D. Cadle, W. H. Fischer, E. R. Frank, and J. P. Lodge, Jr., *J. Atmos. Sci., 25,* 100 (1968); R. W. Fenn, H. E. Gerber, and D. Wasshausen, *ibid., 20,* 466 (1963).

3
C. E. Junge, *Air Chemistry and Radioactivity* (New York: Academic Press, 1963).

4
R. D. Cadle and M. Ledford, *Int. J. Air Water Pollut., 10,* 25 (1966); J. M. Hales, diss., University of Michigan (1968).

5
G. Liuti, S. Dondes, and P. Harteck, *J. Amer. Chem. Soc., 88,* 3212 (1966).

6
R. D. Cadle, in *Air Pollution Handbook,* P. L. Magill, F. R. Holden, and C. Ackley, eds. (New York: McGraw-Hill, 1956), pp. 3-1 to 3-27.

7
E. R. Gerhard and H. F. Johnstone, *Ind. Eng. Chem., 47,* 972 (1955); T. C. Hall, Jr., diss., University of California, Los Angeles (1963); N. A. Renzetti and G. T. Doyle, *J. Air Pollut. Control Assoc., 8,* 293 (1959); *Int. J. Air Pollut., 2,* 327 (1960).

8
P. Urone, H. Lutsep, C. M. Noyes, and J. F. Parcher, *Environ. Sci. Technol., 2,* 611 (1968).

9
H. F. Johnstone and D. R. Coughanowr, *Ind. Eng. Chem.*, *50*, 1169 (1958).
10
R. D. Cadle and J. W. Powers, *Tellus*, 18, 176 (1966).
11
R. D. Cadle and R. C. Robbins, *Discuss. Faraday Soc.*, *30*, 155 (1960).
12
E. L. Wong and A. E. Potter, Jr., *J. Chem. Phys.*, *39*, 2211 (1963).
13
R. D. Cadle, *Discuss. Faraday Soc.*, *37*, 66 (1964).
14
H. W. Ford, G. J. Doyle, and N. Endow, *J. Chem. Phys.*, *26*, 1337 (1957).
15
J. P. Lodge, Jr., and J. B. Pate, *Science*, *153*, 408 (1966).
16
H. W. Ford and N. Endow, *J. Chem. Phys.*, *27*, 1156 (1957).
17
F. Kaufman and J. R. Kelso, *ibid.*, *40*, 1162 (1964).
18
H. W. Ford, G. J. Doyle, and N. Endow, *ibid.*, *26*, 1336 (1957).
19
L.R.J. McHaney, diss., University of Illinois (1953).
20
L. A. Cavanagh, C. F. Schadt, and E. Robinson, *Environ. Sci. Technol.*, *3*, 251 (1969).
21
P. A. Leighton, *Photochemistry of Air Pollution* (New York: Academic Press, 1961).
22
R. D. Cadle and H. S. Johnston, in *Proceedings of the Second*

National Air Pollution Symposium (Stanford Research Institute, Menlo Park, California, 1952), pp. 28–34.

23

R. D. Cadle and J. W. Powers, *J. Phys. Chem., 71,* 1702 (1967).

24

W. E. Wilson, Jr., and J. T. O'Donovan, *J. Chem. Phys., 47,* 5455 (1967).

25

M. A. A. Clyne and B. A. Thrush, *Proc. Roy. Soc. London, Ser. A, Math. Phys. Sci., 275,* 559 (1963).

26

J. N. Pitts, Jr., A. U. Khan, E. B. Smith, and R. P. Wayne, *Environ. Sci. Technol., 3,* 241 (1969).

27

I. M. Campbell and B. A. Thrush, *Proc. Roy. Soc. London, Ser. A, Math. Phys. Sci., 296,* 222 (1967); F. Kaufman and J. R. Kelso, *Discuss. Faraday Soc., 37,* 26 (1964); O. R. Lundell, R. D. Ketcheson, H. I. Schiff, paper presented at the 12th International Symposium on Combustion, Poitiers, France, 1968.

28

L. F. Phillips and H. I. Schiff, *J. Chem. Phys. 37,* 1233 (1962).

29

F. Kaufman, *Ann. Geophys. 20,* 106 (1964).

30

R. D. Cadle, R. Bleck, J. P. Shedlovsky, I. H. Blifford, J. Rosinski, and A. L. Lazrus, *J. Appl. Meteorol., 8,* 348 (1969).

31

T. G. Scholz, L. E. Heidt, E. A. Martell, and D. H. Ehalt, *Trans. Amer. Geophys. Union, 50,* 176 (1969).

32

A. E. Bainbridge and L. E. Heidt, *Tellus, 18,* 221 (1966).

33

T. G. Kyle, D. G. Murcray, F. H. Murcray, and W. J. Williams, *J. Geophys. Res., 74,* 3421 (1969).

34

Based on *U.S. Standard Atmosphere 1962* (Washington: U.S. Government Printing Office, 1962).

35

P. M. Furukawa, P. L. Haagenson, and M. J. Scharberg, *Nat. Center Atmos. Res. Tech. Note 26* (1967), pp. 1–55.

36

B. G. Hunt, *J. Geophys. Res., 71,* 1385 (1966).

37

J. B. Pearce, *ibid., 74,* 853 (1969).

16

The Fate of SO_2 and NO_x in the Atmosphere　　　Erik Eriksson

It was, of course, realized long ago that fossil-fuel combustion introduces various substances into the atmosphere. It is also realized that some of them have been accumulating ever since the time of industrial revolution. The accumulation is most obvious for carbon dioxide, a substance of considerable interest, since its long-wave radiation absorption can possibly influence the earth's climate. (See Chapter 14 for a discussion of carbon dioxide in the atmosphere.) The present rate of increase seems to be around 0.6 part per million per year (ppm/yr) in the atmosphere.[1]

As for other substances released or formed through combustion, carbon monoxide should be mentioned since it is a surprisingly inert gas. Recent investigations seem to indicate a complete destruction of this gas in the stratosphere;[2] hence its half-life in the atmosphere has an upper limit of about ten years. If its production rate is known, there should thus be no difficulty in predicting its content in the atmosphere.

There are two other constituents in smoke effluents which are of considerable interest: sulfur dioxide and various nitrogen oxides. As for sulfur dioxide, its source is sulfur in fossil fuel originally contained in the vegetable matter that presumably is also the origin of oil. Its content in coal and oil varies, being as high as 4%–5%. During combustion it is oxidized to sulfur dioxide and released as such. Its further fate in the atmosphere is known fairly well. It is oxidized to sulfur trioxide, which is very hygroscopic and forms sulfuric acid, being added to all kinds of particles in the atmosphere, including cloud drops. As such it is prone to removal by precipitation; particulate matter has a resi-

The references for Chapter 16 are on page 301.

dence time in the atmosphere which barely exceeds a few weeks. The oxidation of sulfur dioxide is enhanced by the presence of alkaline materials such as ammonia gas, calcium carbonate particles, and other soil particles of an alkaline nature in the atmosphere. It is also photochemically oxidized at a rate that would put an upper limit on its residence time in the atmosphere at two to three weeks.

In addition, sulfur dioxide is absorbed by plants and, most likely, by seawater so that the earth's surface constitutes a sink for this gas. Thus sulfur released by fossil-fuel combustion hardly spends more than a few weeks in the atmosphere. Consequently, one cannot expect any steady accumulation of sulfur compounds in the atmosphere. Since there is a "natural" circulation of sulfur in nature, including the atmosphere, an increase in sulfur compounds in the atmosphere should be proportional to the rate of increase in fossil-fuel combustion, not to the amounts released in the past. This seems to be the case.

Figure 16.1 summarizes the circulation of sulfur in nature, including that released by fossil-fuel combustion.[3] It is seen that by far the largest amount of "readily available" sulfur is found in the oceans. Land plants and soils also contain a considerable amount. At the time the diagram was constructed (1962), the figure for sulfur released by human activity was 40 million metric tons per year. More recent estimates have doubled this figure, which means that the amount is about as large as that normally passing through the terrestrial biosphere. However, considering that most of this is released in the northern hemisphere in a rather narrow latitude belt, the proportion of fossil-fuel sulfur will certainly dominate in this belt.

Sulfur dioxide in the atmosphere has been measured in many places but almost exclusively at ground level and in or near urban places. Such figures are of very little value for judging global conditions. Sulfur in precipitation, as sulfate, is a much better measure since it integrates the atmosphere vertically to a con-

16 The Fate of SO_2 and NO_x in the Atmosphere

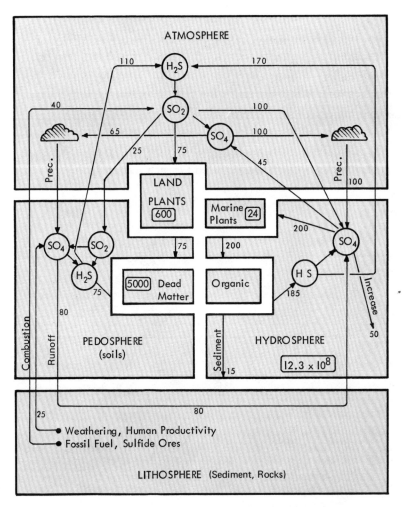

16.1
Circulation of sulfur in nature. The units are millions of metric tons of sulfur. The enclosed figures are simply amounts; the other figures are amount per year.

siderable extent. Data on sulfate in precipitation in Western Europe are available for a period from 1955 up to the present.

These data were recently studied by the author with respect to long-term trends.[4] A map of the location of stations is shown in Figure 16.2. Stations were then grouped according to countries, and in Scandinavia a representative group was made up of stations in the southern part (south Norway, south and middle Sweden). Yearly means were computed, and in order to make stations comparable the following procedure was applied. First I computed the logarithms of yearly mean concentrations for each station (denoted by $\log c$) and then the mean of these values for all the years for each station (denoted by $\overline{\log c}$). The deviations $\log c - \overline{\log c}$ were plotted for each station against

16.2
Location of stations in the West European Atmospheric Chemistry Network.

16 The Fate of SO_2 and NO_x in the Atmosphere

time. The result for the south Scandinavia stations is shown in Figure 16.3. It is apparent from this that there are considerable climatic fluctuations depending on the prevailing atmospheric circulation pattern for each year. It is, however, gratifying to

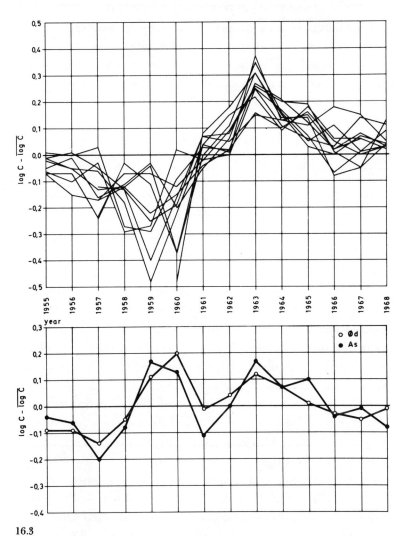

16.3

Log $c - \overline{\log c}$ on a yearly basis for sulfur in precipitation at south Scandinavian stations (upper curves) and Danish stations (lower curves).

find how closely different stations in a region follow one another. It is obvious that it is rather difficult to determine long-term trends from a short series of such observations, but the figure does suggest an average increase of about 2% per year. This is roughly what one might expect from the rate of increase in affluence in Europe during this period, considering the presence of a certain basic "natural" part of the precipitated sulfur. The trend is less obvious on the British Isles where, for example, the northwest part even shows a decrease with time (see Figure 16.4). This is, of course, understandable because the region is on the very border of the European industrial source area for sulfur dioxide. Taking all the countries on the same graph, we see the result shown in Figure 16.5, which, despite the spread of data, still suggests a slow increase with time.

If sulfur should accumulate for a very long time in the atmosphere, say indefinitely, the patterns in the figures shown would be quite unlikely. The rather large amplitude in the climatic fluctuations definitely supports a relatively short atmospheric residence time. Hence, from the point of view of the atmosphere and its physical properties, no substantial global effect can be expected in the future from sulfur dioxide released by fossil-fuel combustion. This does not, however, imply that the sulfur dioxide released is harmless. If sulfur does not accumulate in the atmosphere, it will accumulate somewhere else, and the first area to be affected is naturally the soil.

As for nitrogen oxides, these are mainly formed in combustion processes by reaction between nitrogen and oxygen at high temperatures. This is a well-known reaction and has even been used extensively for nitrogen fertilizer production in the Birkeland-Eyde method. Some ammonia released during combustion may also be oxidized, but the direct "fixation" of atmospheric nitrogen is probably the most important process, enhanced with time through much more efficient combustion techniques. The nitrogen oxide so formed will be oxidized to nitrogen dioxide, NO_2,

16 The Fate of SO₂ and NOₓ in the Atmosphere

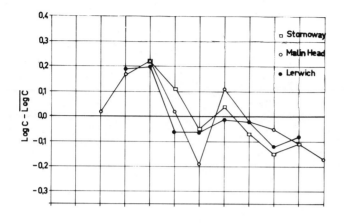

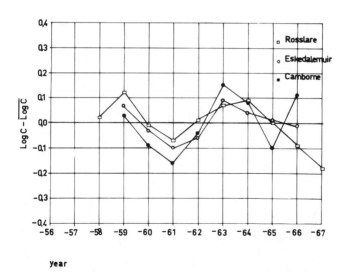

16.4

Log $c - \overline{\log c}$ on a yearly basis for sulfur in precipitation at stations in the British Isles.

which is a constituent normally found in the atmosphere. Nitrogen dioxide is, however, not a new compound in the atmosphere. Since it forms nitrates that are found in precipitation, its presence was detected early in the nineteenth century, and great

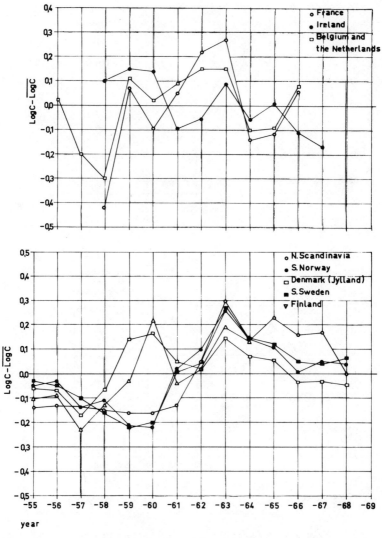

16.5

Log $c - \overline{\log c}$ on a yearly basis for sulfur in precipitation for various countries.

16 The Fate of SO$_2$ and NO$_x$ in the Atmosphere

attention was paid to nitrates in precipitation because of their fertilizing value for terrestrial vegetation. Much has been published on the subject of the origin of nitrates in the atmosphere,[5] with thunderstorm activity being a favorite idea (although never clearly verified). Silent electrical discharges also were thought to produce nitrogen dioxides, and oxidation in the atmosphere of ammonia released from soil has been suggested. Nitrogen is a vital constituent in living matter. Plants prefer nitrates in their nutrient uptake but release the nitrogen as ammonia and as nitrous oxide, a rather inert gas known better as laughing gas.

Precipitation has been analyzed for nitrates in many places over the earth. One of the longest series, beginning about 1900 and covering more than twenty years, originates from Rothamsted in England,[6] where a clear positive trend in nitrate concentration over time was noted. Already the trend was explained as owing to an increased rate of fixation of nitrogen in combustion, partly because of increased fuel consumption and partly because of increased efficiency of combustion. Fixation of nitrogen as nitrogen oxides is known to take place in various internal-combustion engines and is a severe problem in cities with heavy automobile traffic.

As to the atmospheric residence time of nitrogen oxides, nothing suggests a steady accumulation, the residence time being about the same for sulfur compounds. In the Western European network of atmospheric chemistry stations, a clear positive trend is seen also for nitrates, of about the same magnitude as that for sulfate (Figure 16.6). Hence, from the point of view of the physical properties of the atmosphere, one can hardly expect any influence in the future. But nitrates may influence soils and plants.

Sulfur dioxide and nitrogen oxides have one thing in common: their oxidation will turn them into strong acids. Normally, the atmosphere carries enough alkaline material from arid and semi-arid regions to neutralize this acid. In the early industrial era, further alkaline material was added to the atmosphere in ashes

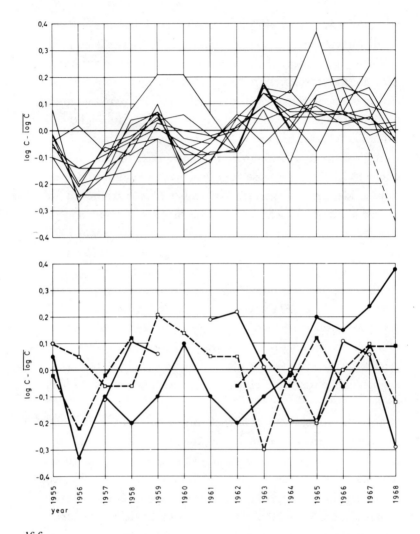

16.6

Log $c - \overline{\log c}$ on a yearly basis for nitrate in precipitation at south Scandinavian stations (upper curves) and at north Scandinavian stations (lower curves).

16 The Fate of SO_2 and NO_x in the Atmosphere

that were part of the smoke. This has changed rather drastically in recent years. Ashes and other alkaline particulates are today largely removed from stack gases since they constitute an unpleasant urban pollutant. Oil, which is poor in ash, is largely replacing coal as fuel. Hence, fossil-fuel combustion adds more and more acid material to the atmosphere, while the supply of alkaline materials to the atmosphere is about the same as before. The balance between acid and alkaline materials is thus greatly disturbed and will be disturbed even more in the future. This shows up as increased corrosion of metals and has lately also shown some rather widespread effects on lakes in western Sweden and southern Norway. The pH values of some lakes have dropped considerably. In some cases valuable game fish have been exterminated, and salmon are avoiding rivers in south Norway. In western Sweden some lakes have pH values below 4. Soils in this area are poor in alkaline material so that they are likely to be depleted of nutrients in the near future if the present situation continues.

The acidity in precipitation in these areas can be seen from Figure 16.7, which shows excess acidity after proper subtraction of alkalinity during months with high pH values. The values shown are in milligram equivalents per square meter (meq/m^2) for the periods in question. It is seen that in 1960–1965 an average of about 30 meq/m^2 per year was recorded in certain parts, which corresponds to about 1.5 metric tons of sulfuric acid per square kilometer per year. With this rate, some of the forest soils will be completely exhausted of nutrients in about 100 years, and a yearly decline in productivity is expected to occur.

Now, western Sweden is especially sensitive; but if sulfur dioxide is released to the atmosphere in ever-increasing amounts in the future, quite severe and widespread damage to soils will definitely occur. The nitrogen oxides naturally add to the acidity, although they also have some nutritional value.

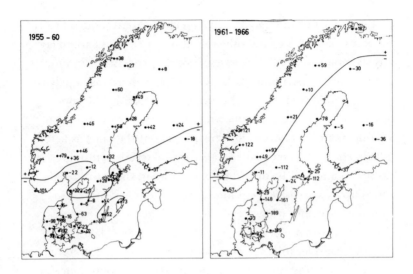
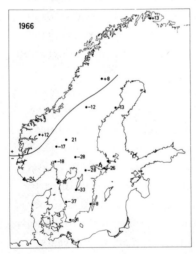

16.7
The sum total of excess base and excess acid in precipitation in Scandinavia for the periods indicated: 1955–1960, 1961–1966, and 1966. The units are milligram equivalents per square meter (meq/m^2). Positive figures indicate excess base; negative figures indicate excess acid.

References

1

B. Bolin and W. Bischof, *Variations of the Carbon Dioxide Content of the Atmosphere,* Report AC-2 (Stockholm: University of Stockholm, Institute of Meteorology, December 1969).

2

W. Seiler and C. Junge, "Decrease of carbon monoxide mixing ratio above the polar tropopause," *Tellus, 21,* 447–449 (1969).

3

E. Ericksson, "The yearly circulation of sulfur in nature," *J. Geophys. Res., 68,* 4001–4008 (1963).

4

E. Ericksson, "The importance of measuring global background pollution," to be published in World Meteorological Organization Technical Note Series.

5

E. Ericksson, "Composition of atmospheric precipitation: I, nitrogen compounds," *Tellus, 3,* 215–232 (1952).

6

E. J. Russel and E. H. Richards, "The amounts and composition of rain and snow falling at Rothamsted," *9,* 309–337.

17

An Isotope-Ratio Method for Tracing Atmospheric Sulfur Pollutants

Meyer Steinberg

Sulfur is one of the major contributors to air pollution in metropolitan areas. It exists primarily in the forms of sulfur dioxide (SO_2) and sulfur trioxide (SO_3), the latter often in combination with particulates as an acid smut or with water as a sulfuric acid mist. The principal source of sulfur oxides in urban areas is fossil fuel (coal and oil), used for power generation, space heating, and industrial processing. The results of sulfur oxide pollution are irritation of the eyes and respiratory system, aggravation of respiratory ailments, plant damage, and corrosion of masonry and metal structures. These effects are enhanced by the conversion of sulfur dioxide to sulfur trioxide.

Despite the importance of sulfur oxides in the overall pollution picture, there are many questions unanswered about their behavior and fate as they are released into the atmosphere. A major reason for this is the lack of a good method of tracing sulfur oxides. Since there are many individual sources, both high-level and low-level, contributing to the usual urban atmosphere, a reliable tracer is necessary to follow the course of sulfur through the air and to identify it with particular sources.

Development of the isotope-ratio tracer method was undertaken to satisfy the need for a means of tracing sulfur that left no doubt as to its disposition. The method makes possible the use of sulfur already in a fuel as a tracer for the sulfur oxides produced by the combustion of that fuel. Thus there is no question of how well the tracer simulates the behavior of the substance being followed, since it is identical with it.

The isotope-ratio tracer method depends upon the fact that there is a natural variation in the ratios of the stable isotopes

The references for Chapter 17 are on pages 315–316.

17 Isotope-Ratio Method for Tracing Sulfur Pollutants

Table 17.1
Isotopic Composition of a Terrestrial Sulfur

S-32	95.10%
S-33	0.74
S-34	4.20
S-36	0.16

making up sulfur from different sources.[1] A typical composition is given in Table 17.1. The variation is the result of isotope fractionation caused by chemical and bacterial action in different environments. The ratio of the two most abundant isotopes, sulfur-32 and sulfur-34, is the measure used to characterize a particular sulfur. A standard ratio, 22.210, has been established based on the ratio in meteoritic sulfur, to which other sulfurs are compared. The difference, in per mil,* is expressed by:

$$\delta(\text{S-34}) = \left[\frac{\frac{(\text{S-32})}{(\text{S-34})} (\text{standard})}{\frac{(\text{S-32})}{(\text{S-34})} (\text{sample})} - 1 \right] \times 1000 \qquad [1]$$

or

$$\delta(\text{S-34}) = \left[\frac{22.210}{\frac{(\text{S-32})}{(\text{S-34})} (\text{sample})} - 1 \right] \times 1000. \qquad [2]$$

Samples having a higher abundance of sulfur-34 than the standard will have a positive $\delta(\text{S-34})$ or a plus "del" value; those having a lower abundance will have a negative $\delta(\text{S-34})$ or a minus "del" value. Table 17.2 shows the general range of "del" values for sulfur from various sources. For any given deposit, the variation in the ratio will be much narrower than the range indicated.

---

*The terminology "per mil" is in frequent usage among those who measure isotopic ratios. The dimensionless unit "per thousand" would be equivalent. Since "del" values are frequently in the range of a few parts per thousand, it is customary to normalize the values to 0.001 and express them in the dimensionless units, "per mil."

Table 17.2
Range of δ(S-34) Values for Sulfur of Different Origins

Coal	+28 to −5 per mil
Oil	+14 to −26
Natural gas	+20 to −3
Gypsum	+35 to +5
Pyrites	+35 to −45
Sulfur deposits	+10 to −10

With oil, there will be little variation within a single pool.

The (S-32)/(S-34) ratio at any point in the atmosphere will be the resultant of the ratios in the sulfur from all the sources contributing SO_2 to that point. If this is relatively constant with time, the use of a fuel having a ratio different from that of the background sulfur would make it possible to trace the SO_2 being emitted from the plant using the fuel.

The sensitivity of the isotope-ratio tracer method depends upon the difference between the isotope ratio in the stack gas and in the background, the variability of the ratio in the background, the amount of SO_2 collected, and the accuracy with which the (S-32)/(S-34) ratio can be measured. With samples of SO_2 weighing at least a milligram, a difference of 10 to 15 per mil between the δ(S-34) values for stack and background sulfur, and a precision of 0.2 per mil in measuring ratios, it is possible to detect about one part of stack SO_2 in the presence of 50 parts of background SO_2.

Utilization of the isotope-ratio tracer method required the development of a sampling system that permitted the collection of adequate size samples of SO_2 (or other sulfur compounds) from the atmosphere in reasonable lengths of time, a reliable chemical separation and purification system to prepare SO_2 for ratio measurement, and an isotope-ratio mass spectrometer capable of performing on small samples.[2]

A system for the collection of SO_2 and SO_3 particulates was developed. In its present form it consists of a glass fiber prefilter that collects particles, either solid or mist, which carry the sulfates, and a chemically treated filter paper that collects SO_2. An 8 by 10 inch sheet of each is placed in a gasketed hinged frame, separated by a stainless steel screen to prevent transfer of sample from one sheet to the other (see Figure 17.1). The filter pack is used in conjunction with a commercially available high-volume air sampler, and the same type of pack is used for both ground-level and airborne sampling. Air flows of greater than 1 cubic meter per minute are possible through the pack. This permits the collection in one hour of the minimum size sample needed for analysis, 1 milligram, from atmospheres having an SO_2 concentration as low as 0.01 part per million (ppm). This would be typical of the lower levels of SO_2 that might be found in a small city.

For sampling aloft, an airscoop assembly was designed that could be fitted into the window frame of a small aircraft (see Figure 17.2). The scoop is equipped with a shutter that permits stopping the air flow to the filter while the plane makes a turn outside of a stack plume that is being traversed.

The measurement of the relative abundance of the sulfur isotopes by the mass spectrometer is made on SO_2 gas. This SO_2 must be of high purity, free of moisture, CO_2, and other gases. Standard methods used in the preparation of SO_2 samples for introduction into mass spectrometers had to be adapted to handle samples as small as 1 milligram. On extraction of the sulfur oxides from the paper filter with water, the sample is subjected to a series of chemical oxidation and reduction steps so that the sulfur can be equilibrated with a standard source of oxygen to eliminate interference effects of oxygen-16 and oxygen-18. The ratio is then performed with a double-beam isotope-ratio mass spectrometer, shown in Figure 17.3. A precision of 0.2 per mil in δ(S-34) is obtained routinely.

There are several ways of obtaining a different sulfur ratio in the stack plume gases from the background sulfur ratio. (1) The sulfur in the power plant may already be different from back-

17.1
Experimental system for air sampling. A glass fiber prefilter collects solid or mist particles that carry sulfates, and a chemically treated filter paper collects SO_2. The filter pack is used with a high volume of air sampler to permit collection of the minimum sample in one hour. The cylindrical canister collects the SF_6.

17 Isotope-Ratio Method for Tracing Sulfur Pollutants

17.2
The high-volume air sampler and instrument package fitted into the window of a Cessna aircraft.

ground. (2) A special fuel may be located having a much different sulfur ratio than background. (3) Sulfur dioxide derived from a high-ratio chemical source, such as calcium sulfate from sea water, may be injected into the stack gases. (4) Finally, if a natural tracer cannot be found, the SO_2 may be isotopically enriched in S-34 by physical means.

A search for special fuel sources indicated that fuel oil would be superior to coal as a sulfur-ratio tracer because the sulfur is homogeneously distributed in the liquid state. The sulfur in coal exists as separated pyrite and as organic sulfur, and there can be a high degree of sulfur heterogeneity. A survey was made of crude

17.3
Double-beam isotope-ratio mass spectrometer.

oil deposits, and several deposits indicated δ(S-34) values varying from 5 to 10 per mil units different from background values measured in the urban areas investigated in this study. (See Table 17.3.) The background values in these areas averaged between 0 and 5 per mil units. Fuel oils obtained from refineries distilling these special crudes would then be adequate tracer sources.

In conjunction with pure tracer studies, it is possible to use the isotope-ratio technique to determine the conversion of SO_2 to SO_3 and sulfates in the power plant plume. As SO_2 is oxidized to the higher oxides, it becomes enriched in the sulfur-34 isotope through the exchange reactions shown in Table 17.4. Equi-

Table 17.3
Isotope Ratio and Sulfur Content of Selected Oils

Source and description	δ(S-34) per mil	Sulfur content
1. River Bend, Wyoming, crude, #PC-66-39	−6.6	2.52%
2. Reno, Wyoming, crude, #PC-66-48	−4.5	0.77
3. Woodrow, Montana, crude, #PC-58-399	−4.8	0.90
4. Sho-Vel-Tum, Oklahoma, crude, #59171	+14.5	1.44
5. Neafra, Kuwait, crude, #67065	−9.8	3.91
6. Safoniya, Saudi Arabia, crude, #67066	−8.5	2.97

librium constants have been calculated for this isotopic enrichment and exchange,[3] and the conversion of SO_2 to higher oxides can be determined by measuring the change in $\delta(S\text{-}34)$ in the SO_2 of the plume at various distances from the stack.

The equilibrium exchange constant K for Reaction 4 in Table 17.4 is related to the quantities of each reactant:

$$K = \frac{(S^{32}O_2)(S^{34}O_4^=)}{(S^{34}O_2)(S^{32}O_4^=)}. \qquad [3]$$

The right-hand side of Equation 3 can be expressed as the quotient of the isotopic ratios of the two different chemical species present in the reaction:

$$R(SO_2) = \frac{(S^{32}O_2)}{(S^{34}O_2)}, \quad R(SO_4^=) = \frac{(S^{32}O_4^=)}{(S^{34}O_4^=)}; \qquad [4]$$

$$K = \frac{R(SO_2)}{R(SO_4^=)}. \qquad [5]$$

Recalling the definition of the "del" value for isotope ratios in Equations 1 and 2, the exchange constant can be expressed as

$$K = \frac{1 + \frac{\delta(SO_4^=)}{1000}}{1 + \frac{\delta(SO_2)}{1000}}. \qquad [6]$$

Table 17.4
Isotope-Ratio Tracer Method for Measuring Conversion of Lower Oxides of Sulfur to Higher Oxides

Oxidation and exchange reactions		Temp. (°C)	Exchange constant K
(1)	$SO_2 + \tfrac{1}{2}O_2 \rightleftharpoons SO_3$		
(2)	$S^{34}O_2 + S^{32}O_3 \rightleftharpoons S^{32}O_2 + S^{34}O_3$	25	1.037
		900	1.003
		1500	1.001
(3)	$S^{34}O_2(g) + HS^{32}O_3^- \text{(Sol'n)}$ $\rightleftharpoons S^{32}O_2(g) + HS^{34}O_3^- \text{(Sol'n)}$	25	1.019
(4)	$S^{34}O_2(g) + S^{32}O_4^= \text{(Sol'n)}$ $\rightleftharpoons S^{32}O_2(g) + S^{34}O_4^= \text{(Sol'n)}$	25	1.041

If none of the SO_2 is converted to SO_3 or other oxidized forms, the "del" value for the SO_2 in the plume, corrected for the normal background of SO_2, would be equal to the "del" value for the fuel consumed in the power station:

$$\frac{\delta(SO_2)}{1000} = \frac{\delta(\text{fuel})}{1000} \quad \text{(no conversion of } SO_2 \text{ to higher oxides)}. \qquad [7]$$

However, conversion does occur, and there is subsequent exchange among the isotopic forms of the various chemical species. This will change the "del" value for SO_2 in the plume. Under the assumption that all of the SO_2 is converted to SO_3, which then hydrates to the form $SO_4^=$, the isotope ratio for the $SO_4^=$ will be equal to the isotope ratio of the fuel. As the isotopic constituents subsequently adjust to the equilibrium of Reaction 4 in Table 17.4, the exchange constant becomes

$$K = \frac{R(SO_2)}{R(\text{fuel})}. \qquad [8]$$

Equation 8, rewritten in terms of the "del" values corresponding to the isotope ratios, can be manipulated into an expression for $\delta(SO_2)$. In the approximation that $K \approx 1$,

$$\frac{\delta(SO_2)}{1000} = \frac{\delta(\text{fuel})}{1000} - (K - 1) \qquad [9]$$

(total initial conversion of SO_2 to the higher oxides).

17 Isotope-Ratio Method for Tracing Sulfur Pollutants

Equations 7 and 9 represent the extremes of no conversion of SO_2 to the higher oxides and total initial conversion to the higher oxides. If only a fraction X of the SO_2 is converted to higher oxides, then an expression can be constructed which satisfies the extreme cases and represents a linear relationship in between:

$$\frac{\delta(SO_2)}{1000} = \frac{\delta(\text{fuel})}{1000} - (K-1)X, \qquad [10a]$$

$$\delta(SO_2) = \delta(\text{fuel}) - 1000(K-1)X. \qquad [10b]$$

A more complete discussion of Equation 10b, including the effects of the additional branching reactions not considered here, can be found in the references.[2,4]

Several meteorological experiments have been undertaken using the isotope-ratio tracer method in the following locations:

1.
In New Haven (Figure 17.4) there is one power plant source and a number of low-level industrial sources of SO_2. For this situation, it has been possible to prepare a mathematical model of air pollution patterns depending on meteorological conditions and sulfur content of fuel. It is anticipated that, once this has been demonstrated for New Haven, it should be feasible to extrapolate the method to other cities without too much difficulty.

2.
Studies at the Keystone plant in Pennsylvania (Figure 17.5), a coal-burning plant with two 800-foot stacks in open country with low SO_2 background, allow the possibility of tracing the plume and determining the fate of SO_2 for many miles across the countryside.

3.
In Northport, Long Island (Figure 17.6), there is an oil-burning plant with a tall stack which can be traced over water as well as over land with little contribution from background.

4.
Finally, it is intended to study a tall stack in a larger multisource

17.4
The English Station (United Illuminating Company) in New Haven, Connecticut, looking south to Long Island Sound. There are a number of sources of SO_2 in an urban region, in addition to power plants.

area such as New York City. For this purpose the Consolidated Edison oil-burning plant at Arthur Kill, Staten Island, will be investigated.

Some interesting preliminary data have been obtained from the Long Island Lighting Company (LILCO) Northport plume, shown in Figure 17.7. In this run, sulfur hexafluoride (SF_6) was injected into the plume as an inert tracer. The simultaneous use of tracer SF_6 and sulfur-ratio analysis makes it possible to correct for dilution of the plume SO_2 with background SO_2 and to determine the divergence of chemically active and inactive species. The calculated conversion of SO_2 to SO_3 along the plume is shown in Table 17.5. The data indicate that there is an initial formation of SO_3 in the power plant. In the first kilometer about

17.6
The Northport plant (Long Island Lighting Company) on Long Island Sound. The results of measurements taken in the plume of this power plant are presented in Table 17.5 and Figure 17.7.

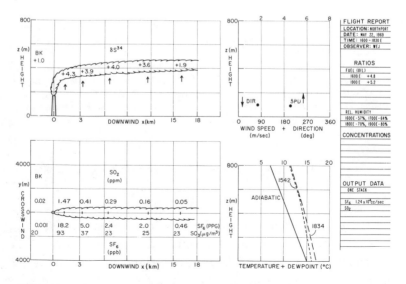

17.7
Experimental observations made in the plume of the Northport power plant.

17 Isotope-Ratio Method for Tracing Sulfur Pollutants

17.5
The Keystone Generating Station in Shelocta, Pennsylvania. This station is approximately 40 miles from Pittsburgh and is located in gently rolling farmland. It burns 650 tons of bituminous coal per hour which is delivered by conveyor belt and truck from nearby mines. The stacks are 800 feet tall, and there are four natural-draft cooling towers. (The station is owned jointly by Atlantic City Electric Company, Baltimore Gas and Electric Company, Delmarva Power and Light Company, Jersey Central Power and Light Company, Pennsylvania Power and Light Company, Philadelphia Electric Company, and Public Service Electric and Gas Company.)

1.5% conversion takes place which slowly rises to 2.4% at 12 km; then a more rapid rise to 4.5% at 15 km is found. Thus, as the stack gases enter the atmosphere and are rapidly diluted by several orders of magnitude with cool moist air, oxidation will take place, either directly or more rapidly after SO_2 is dissolved in condensed water droplets. Further downwind there is a slower oxidation as shown by the slow decrease in $\delta(S\text{-}34)$ value. Finally, at a large distance the conversion increases again. The isotope-

Table 17.5
Northport Power Plant, May 22, 1969

Distance from stack (km)	SF$_6$ concentration ppb by volume	SO$_2$ concentration ppm by volume	δ(S-34) per mil measured	δ(S-34) per mil plume	X fraction SO$_2$ converted
1.3	18.2	1.47	+4.3	+4.3	0.015 ± .005
3.2	5.0	0.41	+3.9	+4.2	0.017 ± .005
6.5	2.4	0.29	+4.0	+4.0	0.022 ± .005
11.5	2.0	0.16	+3.9	+3.9	0.024 ± .005
16.0	0.46	0.047	+1.9	+3.0	0.046 ± .005

Note. The background concentration of SO$_2$ was 0.024 ppm. The "del" value for the background was δ(S-34) = +1.0 per mil. The "del" value for the sulfur in the fuel was δ(fuel) = +4.9 per mil. This value was selected for the calculation; "del" values for the fuel burned varied from 4.8 to 5.2 during the time of measurement.

ratio tracer method of determining SO$_2$ conversion is independent of the size of sample taken and thus is much more sensitive and precise than the direct SO$_2$ measurement. It is estimated that the isotope-ratio tracer method can give a much more precise measurement of the conversion of SO$_2$ to SO$_3$ (3% error for this method versus 20% for direct SO$_2$ measurement).

There is also evidence from ground-level samples using the isotope-ratio tracer method that the conversion is a function of the humidity in the atmosphere: the higher the humidity, the greater the conversion.

It is hoped that, with a more precise understanding of the fate of SO$_2$ around power plant locations, a suitable effective method of control can be developed.

References

1
K. Randoma, *Isotope Geology* (New York: Academic Press, 1954), pp. 276–289.

2
B. Manowitz, B. Smith, M. Steinberg, and W. Tucker, *Status*

Report on the BNL Atmospheric Diagnostics Program, Report BNL 11465 (Upton, New York: Brookhaven National Laboratory, June 1967).

3

H. G. Thode, J. Monster, and H. B. Dunford, "Sulphur isotope geochemistry," *Geochimica et cosmo-chimica acta, 25,* 154–174 (1961).

4

B. Manowitz, W. D. Tucker, R. D. Baldwin, L. A. Cohen, J. Forrest, L. Newman, M. E. Smith, and M. Steinberg, *The Atmospheric Diagnostics Program at Brookhaven National Laboratory: Second Status Report,* Report BNL 50206 (T-553) (Upton, New York: Brookhaven National Laboratory, November 1969), available from Clearinghouse for Federal Scientific and Technical Information, National Bureau of Standards, Springfield, Virginia, 22151.

18

Environmental Aspects   Harry Perry
of Coal Mining

The future for coal in the United States has never looked brighter. With the great increases in demands on energy that have been projected, total coal consumption should continue to increase at least to the end of the century. With coal resources representing about 80% of total U.S. energy resources, sufficient coal is available to meet the challenge that an expanding economy will place on energy supply.

Although much of the projected increase in coal consumption will be at electric generating plants, there will be increases in the demand for higher-quality coals by the steel and other industries at home and abroad. However, with the growing public concern about the environment, electric utilities will have to enlarge greatly their generating capacity and still not increase their pollution loads. As a result, the specifications for coal that they purchase will have to be modified so as to reduce particulate and sulfur dioxide emissions, unless alternate control measures can be found. If a change in specification occurs, the utilities will be in direct competition with steel companies, exporters, and other coal users for a higher-quality coal than they have historically used. This could have an important impact on what coal seams are mined and where the mining is done.

The amount of coal that is surface or strip mined has been rising steadily for many years, increasing from 100 million tons in 1945 to 186 million tons in 1968, and it now represents about one-third of total production. Since productivity at strip mines is about twice that of underground mines and costs of coal produced by stripping are significantly less than coal produced underground, this trend can be expected to continue if all other factors

General reference works on the subject matter of Chapter 18 are listed on pages 338–339.

remain constant. However, much of the strip-mined coal in the eastern part of the United States, particularly, is high in sulfur even after intensive coal preparation, and the reserves of strippable coal are more limited than those that can be mined by underground methods. Changing technological factors may also be important. Among these are relative rates at which strip and underground mining techniques are improved through research and development and what procedure emerges as the lowest-cost method of sulfur dioxide control at power plants. For the immediate future, however, a continuation of growth in strip mining must be expected.

Coal Mining and the Environment
Coal mining has multiple adverse impacts on the environment: disturbance of the land by strip mining with its attendant effect on streams and adverse impact on aesthetic values; acid mine drainage from both strip mines and underground mines; subsidence of the land as a result of underground mining; mine fires occurring mainly in underground mines; and the effect on the land, water, and air of refuse banks created by mining and coal preparation. Of these five environmental problems, damages from strip mining and from acid mine drainage are the most extensive.

There are two methods of strip mining. Area strip mining is done in relatively flat areas (see Figure 18.1). By this method, overburden is removed and piled alongside the "cut" until the coal seam is reached. After coal is extracted, the overburden parallel to this first cut is removed and deposited into the area of the first cut. This operation is repeated until the property line is reached. The land mined by this method leaves ridges and valleys like a washboard unless corrective action is taken.

Contour mining (Figure 18.2) is practiced in mountainous country. At the coal outcrop in the hillside, the overburden is removed by a first cut directly above the coal and then coal is re-

18 Environmental Aspects of Coal Mining

18.1
Area strip mining in western Kentucky. Large shovels remove the overburden to expose a seam of coal. When the coal is exhausted, it is covered with the overburden from the next cut. The result is ugly, unproductive ridges of rock, shale, slag, and mineral waste. Photo courtesy of Billy Davis, *Louisville Courier-Journal*.

covered. Each successive cut removes more overburden, since the coal seam is covered by an increasing amount of overburden as the mining moves from the outcrop toward the center of the hill. Mining follows the contour of the hillside and continues until the proportion of overburden to coal seam thickness makes it uneconomic to mine. In this type of strip mining the spoil or overburden is disposed of by casting it down the mountainside below the coal seam. Unless the discard is stabilized, severe erosion and landslides may result, and the flora and fauna below the bench where coal has been extracted can be damaged. Strip mining also

18.2
Contour strip mining for coal on the mountain slopes of Anderson County, Tennessee. The overburden is removed and cast down the lower slope to expose a seam of coal. Photo courtesy of James A. Curry, Tennessee Valley Authority.

adversely affects streams through the acid mine water that is created and through silting of the streams if the spoil banks are not stabilized.

In any consideration of the effect of coal mining on the environment, the problem must be divided into a consideration of the damages caused by past mining operations and of the prevention of such damage in the future (see Figure 18.3). Correction of past damages may require actions that could be avoided in the future if the mining method used takes reclamation of the land into consideration when the coal is first mined. Moreover, the

18 Environmental Aspects of Coal Mining

18.3
Wasted, eroded land remaining after strip mining in western Kentucky. Corrective action should become part of the mining operation. Photo courtesy of Billy Davis, *Louisville Courier-Journal*.

responsibility for past damages may be difficult to ascribe, but this is not the case for present or future mining.

Subsidence
Subsidence results when less coal is left in place than is necessary to support the overburden. It is of major importance in urban areas, although uneven subsidence in rural areas can prevent the use of the land for crop farming, damage drainage channels, and prevent the construction of such surface facilities as dams, reservoirs, and canals. In urban areas, the settling of rock and earth can cause damage to homes, streets, buildings and other surface structures (see Figure 18.4) and in some instances can endanger the safety of persons living in these areas. The costs of damage from subsidence in urban areas, such as Scranton and Wilkes-Barre, Pennsylvania, can be very high.

The amount of subsidence and the time at which it occurs after mining is a function of how much coal has been left in place to support the overburden, the depth and type of the overburden, the thickness of the coal seam, and the degree of exposure of the coal pillars to factors in the mine environment that cause the coal of the pillar to spall and become reduced in size.

18.4
Subsidence damage in urban areas of eastern Pennsylvania results in street cave-in and destruction of dwellings.

18 Environmental Aspects of Coal Mining

Since the start of mining, the total coal that has been extracted by underground mining and lost during mining is estimated at approximately 69 billion tons. If this coal averaged 5 feet in thickness, then about 7 million acres have been undermined, of which about one-third has subsided. Total production between 1970 and the year 2000 is estimated at 20 billion tons. If 60% is from underground mines, then 1.2 million additional acres would be involved.

Mine Fires

Mine fires are generally found in abandoned coal mines or along the outcrop of coal seams. They can result from an uncontrolled fire that caused the mine to be abandoned or from either natural or man-made fires started near the entrances of an abandoned mine or near the outcrop of the seam. The heat from mine fires can destroy vegetation and, by burning the pillars that have been left to support the surface, can cause additional and unexpected subsidence. In some instances fumes from these fires reach the surface and, when they contain poisonous gases, may endanger human lives. Although since 1949 some work has been done by the U.S. Bureau of Mines in coal mine fire control and 150 fires have been controlled, there are at least 200 additional fires that are destroying valuable natural resources and may be endangering life and property. If we are to protect the environment adequately, these fires and others that may be started must be controlled. The costs of fire control are high, but so are the damages they cause.

Refuse Banks

Underground mining of coal produces waste material that is brought to the surface with the coal and is usually disposed of on the surface (see Figure 18.5). Both strip coal and coal mined in underground operations produce a waste material when cleaned in coal preparation plants. Since 1930, when less than 10% of bi-

18.5
Hazelton Shaft refuse banks adjacent to an urban section of Hazelton, Pennsylvania.

tuminous coal was washed, the percentage of bituminous coal production subjected to some form of mechanical cleaning has increased steadily to about 65% in the early 1970s. Since then, the percentage of bituminous coal prepared has remained relatively constant. The percentage of waste in each ton of coal produced has also increased from about 10% in 1930 to about 22% in 1967. A total of about 1.3 billion tons of bituminous rejects from cleaning plants have been produced to date. An additional unestimated quantity of rejects from hand preparation has been produced. If this refuse has an average density of approximately 1 ton per cubic yard and the average height of the refuse banks is 75 feet, then the waste material from bituminous coal preparation plants would cover 11,000 acres. An additional 12,000 acres are covered with refuse resulting from anthracite mining, and this is concentrated in a relatively small geographic area in eastern Pennsylvania.

Assuming that the same percentage of coal is cleaned as at present (65%) and that the amount of rejected material remains the same (22%), then between 1969 and 2000 an additional 2.5 billion tons of rejects would be generated. These could, if piled to a height of 75 feet, occupy an additional 21,000 acres of land.

The nature of the refuse in these banks varies from location to location depending on whether the material was a reject from the preparation plant or refuse from hand-picked coal. The chemical and physical properties of refuse banks also vary depending on the seam from which the coal was produced and the nature of the strata associated with the coal.

Most of the refuse banks are in rural areas, but those in urban areas are especially objectionable aesthetically (see Figures 18.5 and 18.6). In a survey conducted in 1963, 495 refuse banks in 15 states were found to be on fire, causing local air pollution problems (see Figures 18.7 and 18.8). No survey has been made, but many miles of streams are polluted through leaching of refuse piles by rain water.

The most desirable solution to the refuse problem would be to find uses for this waste material, particularly those banks in urban areas or those in rural locations that are causing the most severe environmental problems. Possible uses are in the manufacture of lightweight aggregates for concrete or cinder blocks, as a fill for flat coastal areas, for the manufacture of bricks, in the construction of secondary roads in the form of burned refuse, and in generating electricity (see Figure 18.6). In mining areas, the refuse has been used to some extent to backfill both active and abandoned underground mines.

If use for the material cannot be found, research is needed on methods to decorate or conceal the banks by finding vegetation

18.6
The Mineral Springs Refuse Bank in the Parsons section of Wilkes-Barre, Pennsylvania. The UGI Corporation is planning to burn this refuse bank to generate electrical power (see Chapter 12, page 175).

that will grow on the banks or by grading and landscaping the banks so that they will blend into the environment.

Strip Mining
It has been estimated that as of June 1, 1965, about 1.3 million acres have been disturbed by strip mining of coal in the United States. The acreage is about equally divided between contour and area mining. Seventy percent of the land affected occurs in the five states of Pennsylvania, Ohio, West Virginia, Illinois, and Kentucky, and over 95% is in private ownership. Unlike some of the other minerals that are surface mined, there is an in-

18 Environmental Aspects of Coal Mining

18.7
The Baker Bank; coarse refuse on fire at Scranton, Pennsylvania.

creasing percentage of coal lands that have been completely reclaimed. Of about 70% of the land disturbed in 1964, approximately 32,000 acres, one-third was partially reclaimed and two-thirds were completely reclaimed. Costs of complete reclamation averaged about $230 per acre and partial reclamation about $150 per acre.

There are over 1,500 bituminous coal and lignite strip mines in the United States, but about 5% of these produced over 50% of the coal mined by stripping. On the other hand, about half of the strip mines produced less than 5% of strip production. Costs of reclamation vary with the type of strip mining and other factors but should not, under most circumstances, be more than 5 to 10 cents per ton of coal.

Strip mining, where adequate reclamation is not undertaken,

18.8
Attempting to quench a burning refuse bank.

can result in adverse affects on aesthetic values, recreational activities, forests, fish and wildlife, land-use potential, and, most important, streams. This occurs when acid mine drainage and silting results from land erosion of unstabilized spoil banks (see Figure 18.3). It is possible, however, to prevent much of the damage that can be caused by strip mining. Basic reclamation involves grading; construction of drainage control dams, diversion ditches and stream channels; pond stabilization; and planting to achieve soil stabilization so as to prevent erosion and slides. If basic reclamation were applied to all of the approximately 800,000 acres of land disturbed by coal mining that need some degree of treatment, the costs would be in the range of $300 million, or about $380 an acre. If only those areas causing the most serious damage were to be treated, the total costs could be reduced to $120 million.

18 Environmental Aspects of Coal Mining

Rehabilitation, following reclamation, involves the development of the land for specialized uses. These include cropland, pasture and rangeland, wildlife, recreation, construction of ponds, occupancy for residential or commercial use, and stream improvement. Each of those end uses requires different degrees of land treatment in addition to basic reclamation. Reclamation plus rehabilitation would result in a cost of $480 million. However, the cost-benefit ratio for rehabilitation is superior to basic reclamation for all disturbed areas and less favorable than for selective reclamation of those areas causing the greatest damage. Figures 18.9, 18.10, and 18.11 are photographs depicting various stages of rehabilitation in strip-mined areas.

Reclamation or rehabilitation can be accomplished for all disturbed lands using knowledge that has already been developed. Some experiments have been conducted by the Bureau of Mines in cooperation with the U.S. Department of Agriculture and state agencies in four counties in Pennsylvania. These studies produced useful information on methods and costs of restoring lands that were stripped for coal. Lower-cost methods for different physical conditions, however, are still to be developed. Fundamental information is especially needed concerning nutrient deficiency of soils, effects of bacterial action, groundwater hydrology, and on the character of waste or spoil bank materials. Applied research is needed on improved mining methods, slope stabilization, erosion control, and prevention of acid mine drainage; low-cost effective methods of planting must be found and a determination is needed of which plants will grow most rapidly and stabilize the ground most effectively on various types of low-quality soils.

Acid Mine Drainage

Strip mining can cause acid mine drainage when inadequate or limited reclamation of the disturbed land is undertaken. However, surveys by the Federal Water Quality Administration and

18.9
Hovering over a West Virginia hillside, a helicopter spreads grass and tree seed to help reclaim contour strip-mined land. The seed drops from a motor-driven spreader on a hopper hung below the helicopter. Such techniques allow rapid large-scale seeding of rugged terrain. Photo courtesy of West Virginia Surface Mining Association, 1969.

others have indicated that a much greater portion of the acid mine drainage comes from underground mines. According to the Appalachian Regional Commission in the Appalachian area where the bulk of the mine drainage occurs, over 70% of the acid mine water originates from underground mines, nearly all of which are abandoned.

18 Environmental Aspects of Coal Mining

18.10
This fishing pond in central Ohio was once a strip mine. The coal was mined, and then the area was transformed into a haven for the outdoorsman. The dense tree growth is five years old; the younger growth is one to three years old. There are six ponds in the area. Photo courtesy of Consolidation Coal Company, 1966.

These mining operations in Appalachia were estimated in 1966 to discharge 4 million tons of acid per year into 5,700 miles of streams, resulting in serious deterioration of water quality on a continuous basis. Nearly 11,000 miles of streams in the United States are affected on either a continuous or an intermittent basis.

18.11
The Fairground State Park, scene of the Hambletonian harness racing classic in DuQuoin, Illinois, is partly built on land reclaimed after strip coal mining. Land in the foreground of this picture, up to the curving road that skirts the race track, was once mined for coal and has been reclaimed as parkland. Photo courtesy of National Coal Association.

Pollution from acid mine drainage is estimated to add $3.5 million annually in costs to industrial water users, municipal water supplies, and navigation and public facilities. To these measurable costs must be added the general environmental degradation, the destruction of aquatic life, and the deterrent to water-based recreation.

Methods used to control acid mine drainage depend on the type of mine — strip mine, underground mine above drainage level, or underground mine below drainage level — whether the mine is active or inactive, and on geologic and hydrologic conditions. In surface mining, control of acid mine drainage is easier than for underground mines above the water table. Burial of spoil and

18 Environmental Aspects of Coal Mining

other pyritic-bearing materials, flooding of toxic materials, diversion of water, rapid removal of water that reaches the mining operation, and revegetation of the mined-out areas are all useful methods to reduce acid mine drainage. Abandoned surface mines, however, present more expensive reclamation problems.

Acid mine drainage from underground mines is more difficult to control than that from surface mines; preventing water from entering the mine and rapid removal of water that does get into the mine are effective methods to reduce pollution. When mines below drainage level are abandoned, flooding reduces acid mine water formation. Flooding abandoned mines above drainage level, after sealing of mine openings by various methods, is also being tested.

The effectiveness of air sealing, which was widely used in attempts made during the 1930s to control acid mine water formation, is still a subject of controversy. Based on recent data at several experimental sites, it appears that there is some reduction in the amount of acid produced when air sealing is used. Part of the difficulty encountered with air sealing is that of closing all of the places where air can enter the mine, since changes in barometric pressure allow the mine to "breathe" if there are unplugged openings.

A listing of the methods that have been proposed and tested, some to a limited extent, are shown in Table 18.1. The methods are divided into two main categories, treatment and at-the-source techniques. The effectiveness of the treatment techniques is much greater than at-the-source methods.

The treatment methods are:

1.
Neutralization: The method of using limestone to neutralize the acid is one of the oldest that has been studied and was even used to a limited extent as early as the 1920s. Because of cost, no general application was made of the method at that time. With the

Table 18.1
Cost and Effectiveness of Various Techniques for Controlling Acid Mine Drainage

Technique	Effectiveness	Cost
Treatment:		
Neutralization	80–97%	$.10–1.30/1000 gal
Distillation	97–99	.40–3.25
Reverse osmosis	90–97	.68–2.57
Ion exchange	90–98	.61–2.53
Freezing	90–99	.67–3.23
Electrodialysis	25–95	.58–2.52
At the source:		
Water diversion	25–75%	$300–2,000/acre
Mine sealing	10–80	$1,000–20,000/seal
Surface restoration	25–75	$300–3,000/acre
Revegetation	5–25	$70–350/acre

enactment of recent laws designed to prevent water pollution from acid mine drainage, there has been a renewed interest in neutralization, and many new studies have been initiated.

2.
Distillation: Flash distillation has been reported to be the most promising of the distillation processes that could be used to treat acid mine water. The major difficulties expected appear to be scaling and corrosion. Disposal of the brine produced could also create problems. A very high quality water is produced, and there is extensive experience and technology being developed for design of large-scale water-desalting plants.

3.
Reverse osmosis: This method of treating acid mine water is also being studied as an extension of research on reverse osmosis to desalt water. Tests on acid mine water have been conducted at several field locations, and the preliminary data indicate that the process may have great potential for concentrating the dissolved solids in mine drainage and producing a high-quality water, low in sulfate, iron, calcium, and acidity.

18 Environmental Aspects of Coal Mining

4.

Ion exchange: This process removes a number of undesirable constituents, such as calcium and sulfate, but must be used in combination with some method of neutralization. A number of methods for using ion exchange for acid mine water have been suggested, and two have been tested using synthetic mine water.

5.

Freezing: This method of separating unwanted solids may be applicable to the treatment of acid mine drainage, but no extensive tests have been made on actual acid mine drainage water.

6.

Electrodialysis: This method has been used in a large-scale plant for treatment of brackish water, and because of the low concentration of salts in acid mine water, electrodialysis may be suitable for its treatment. If pretreatment is done to remove certain contaminants, there should be no major problems in using the method on acid mine water.

The costs of treatment shown in Table 18.1 have a wide range of values for any given method and depend on a large number of factors that vary from mine to mine. The Department of the Interior has estimated that the cost of abating acid mine drainage for the nation would be $6.6 billion. The state of Pennsylvania estimated that it would cost $1–2 billion to clean up acid mine drainage in that state alone. In view of the large total costs involved for abatement and the relatively small direct savings that can be demonstrated as a result of abatement expenditures, it will be necessary to select priority areas for treatment (the benefit-cost ratio for total abatement is unfavorable).

The Appalachian Regional Commission, in cooperation with the National Academy of Sciences, has recently completed a one-year study of acid mine drainage pollution. Among other recommendations the study suggests:

An action program for controlling and abating acid mine drainage should be part of a more comprehensive pollution control and environmental improvement program for the lands and waters in designated watersheds. Prior to embarking upon a comprehensive action program, therefore, the state or states in question should be required to identify specific watersheds in order of priority in which acid mine drainage, together with other environmental conditions, is a major impediment to future social and economic development and to submit such priorities to the Secretary of the Interior.

In Pennsylvania $150 million will be spent over a ten-year period for acid mine water abatement. While this is the largest sum earmarked for this purpose by any single governmental body, it will correct only a small portion of the acid mine problem in that state.

Conclusions

Coal mining has created numerous environmental problems that are largely concentrated in Appalachia. Most of the damage to the environment is being caused by past mining operations and leaves the responsibility for correction of the problem and the provision of required funds to the government. With proper attention to mining methods and control procedures, mining can now be undertaken with only minimal effect on the environment. Costs for preventing pollution will vary from mine to mine, but these must be viewed as part of normal business costs.

Subsidence, mine fires, and refuse banks result from coal mining, and they have adverse environmental effects. However, methods to prevent their occurrence or to control the adverse effects, if occurrence cannot be prevented, are generally available. Additional research should lead to lower-cost methods of prevention or control.

Strip-mining environmental damage from current and future operations can largely be prevented using techniques already developed. Over the past several years a number of states have passed surface mining laws which contain land reclamation standards that, if properly enforced, will prevent future damage from surface mining. Reclamation or rehabilitation of the land disturbed by past surface mining operations will be costly, and in many areas the costs can be expected to exceed the benefits. A public commitment will be needed for those areas where unfavorable benefit-cost ratios are foreseen. In such cases the public will need to decide whether it wishes to spend public funds to achieve the intangible benefits of clean water, clean air, and improved aesthetic quality.

Correction of acid mine water pollution will be the most difficult and expensive of all of the environmental problems caused by coal mining. Despite many years of intermittent and sporadic attempts to find new methods to correct this problem, particularly in abandoned mines, no low-cost solution has been found. However, the first large-scale and comprehensive research effort to develop new low-cost control techniques has been initiated. It is hoped that improved methods can be found to control acid mine water from abandoned mining operations. If successful, this achievement combined with preventive measures taken for current mining operations should provide a way for coal to be mined to supply the growing energy needs of the nation while preventing the damages from acid mine drainage.

Discussion

Question from Robert E. Farmer, Jr., a plant physiologist with the Division of Forestry, Fisheries, and Wildlife Development of the Tennessee Valley Authority.

What long-range uses are most suitable for mountainous strip-mined areas? Could they be managed with profit to society as wildlife habitats after reclamation?

Answer by Harry Perry.

Although the damages resulting from stripping operations in mountainous areas of the country usually do not have a pronounced effect on the urban environment, they can despoil areas that possess scenic beauty and, if not reclaimed, disrupt fish and wildlife.

Techniques and methods are known which will prevent deterioration of fish and wildlife habitat and largely correct previous damage. Based upon the success of state projects to reclaim mined land for fish and wildlife, of the 25 states reporting, only two reported no success with respect to fish and one reported lack of success with wildlife.

References

1

Appalachian Regional Commission, "Acid mine drainage: An expensive problem to solve," *Appalachia* (journal of the Appalachian Regional Commission, Washington, D.C.), 2, No. 10, 20–24 (August 1969).

2

William M. Spaulding and Ronald D. Ogden, U.S. Bureau of Sport Fisheries and Wildlife, *Effects of Surface Mining on the Fish and Wildlife Resources of the United States* (Washington: U.S. Department of the Interior, 1968).

3

Ronald D. Hill, Federal Water Quality Administration, *Mine Drainage Treatment: State of the Art and Research Needs* (Washington: U.S. Department of the Interior, 1968).

4

John C. MacCartney and Ralph H. Waite, *Pennsylvania Anthracite Refuse,* Bureau of Mines Information Circular 8409 (Washington: U.S. Department of the Interior, 1969).

5

U.S. Department of the Interior, *Surface Mining and Our Environment* (Washington: U.S. Government Printing Office, 1967).

6

R. W. Stahl, *Survey of Burning Coal Mine Refuse Banks,* Bureau of Mines Information Circular 8209 (Washington: U.S. Department of the Interior, 1964).

7

Edwin M. Murphy, Malcolm O. Magnuson, Pete Suder, and John

Nagy, *Use of Fly Ash for Remote Filling of Underground Cavities and Passageways,* Bureau of Mines Reports of Investigations 7214 (Washington: U.S. Department of the Interior, 1968).

8

Vernal A. Danielson, *Waste Disposal Costs at Two Coal Mines in Kentucky and Alabama,* Bureau of Mines Information Circular 8406 (Washington: U.S. Department of the Interior, 1969).

V Waste Heat

19

Environmental Quality and the Economics of Cooling S. Fred Singer

There are serious conflicts in our society between the demands for more electric power and the desire to protect our natural environment. These conflicts can be adjusted by available technology and at a reasonable cost. We have the legislation and the leadership in government to assure that this is done.

Activities of the Department of the Interior

The Department of the Interior's activities embrace the wide range of natural resources management — for conservation as well as for development. On the conservation side, the Bureaus of the Fish and Wildlife Service, the National Park Service, and the Bureau of Outdoor Recreation have the principal responsibility at the federal level for fish and wildlife, for the nation's parks and wilderness areas, and for the improvement of the quality, both physically and aesthetically, of the places in which we spend our leisure hours.

The Bureau of Land Management is charged with the conservation and prudent utilization of our vast public land resource. The Bureau of Reclamation develops water resources; the four Regional Power Administrations market electrical power; the Bureau of Mines, the Geological Survey, the Office of Coal Research and associated offices, and the Office of Saline Water — all seek to strengthen the nation's posture in the management of land, water, power, and mineral resources.

These activities have given the department direct involvement and a vast background of experience in virtually all aspects of the electric power industry. They have required the department to consider both the need for additional power capacity and the related environmental problems from the viewpoint of what will be best for society as a whole.

Table 19.1
U.S. Electric Power — Past Use, Future Requirements

Year	Billion kWh
1912	11.6
1960	753
1965	1,060
1970	1,503
1975	2,022
1980	2,754
1985	3,639

Source. Pacific Northwest Water Laboratory, *Industrial Waste Guide on Thermal Pollution*, rev. ed. (Corvallis, Oregon: U.S. Department of the Interior, Federal Water Pollution Control Administration, 1968). The Federal Water Pollution Control Administration is now called the Federal Water Quality Administration.

The Dimensions of Thermal Pollution

Thermal pollution is of major concern because of its burgeoning impact on the environment now and in the foreseeable future. Recent analyses show a doubling of electrical needs in the United States every ten years and, in some areas, as little as every six years. (See Table 19.1.)

The electric power industry uses about 80% of all the water taken in by industry for cooling purposes. Other sources of thermal pollution exist, of course, but their impact upon the environment is only a fraction that of the electric power industry. (See Table 19.2.)

In addition to the increase of waste heat because of the greater amount of power produced, an increase can be expected because of the growing percentage of power produced by thermal, and particularly nuclear,* generating plants. Today hydroelectric plants that do not contribute to thermal pollution account for less than 20% of the power produced. Of the 81% produced by thermal plants, 95% is produced by fossil-fuel plants. As the remain-

---

*For every kilowatt-hour (kWh) of energy produced, 6,000 Btu (about two-thirds of the heat) is wasted in fossil-fuel plants and about 10,000 Btu in present-day nuclear plants (1 Btu = 1,055 joules; 1 kWh = 3,413 Btu).

Table 19.2
Use of Cooling Water by U.S. Industry, 1964

Industry	Cooling-water intake (billions of gallons)	% of total
Electric power	40,680	81.3
Primary metals	3,387	6.8
Chemical and allied products	3,120	6.2
Petroleum and coal products	1,212	2.4
Paper and allied products	607	1.2
Food and kindred products	392	0.8
Machinery	164	0.3
Rubber and plastics	128	0.3
Transportation equipment	102	0.2
All other	273	0.5
Total	50,065	100.0

Source. See Table 19.1.

ing suitable sites for hydroelectric plants are limited, the power industry will depend more and more on thermal power generation to meet power needs. Predictions indicate that in 1990 between 92% and 94% of the power generated will be produced thermally, with almost two-thirds of this produced by nuclear plants that require 40% more cooling water than fossil-fuel plants.

Not only will there be increased waste heat on a nationwide scale, there will also be massive concentrations of heat loads. The average size of all units retired between 1961 and 1965 was 22 megawatts. The average size of the 217 new fossil-fuel units under construction or in the planning stages for operation by 1971 is 295 megawatts. The new nuclear units planned for operation by 1973 average 624 megawatts per unit.

By 1980, electrical needs will require the use of one-sixth of the total available fresh-water runoff in the entire nation for cooling purposes. If we discount flood flows that usually occur about one-third of the year and account for two-thirds of the total runoff, it becomes apparent that the power industry will require about half the total runoff for the remaining two-thirds of the year. More recent projections of the waste heat load by the year 2000, which take into account probable technical improvements, indicate that the requirement for cooling water will equal two-thirds of the total national daily runoff of 1,200 billion gallons.

Effects of Thermal Pollution

The projected cooling-water requirement is for gross usage. Even though some 95% of water used in once-through cooling is returned to the stream, the increased temperature of the water has a potentially deleterious effect on other water uses, either by nature or by man's civilization.

These deleterious effects are numerous, but I will point out only a few here.

1.
The capacity of the stream to assimilate other wastes is decreased, forcing other stream users to increase the quality of their effluents if water quality standards are to be met for the stream as a whole.

2.
The oxygen content of the stream is reduced relative to the cooler water of the natural stream state; this oxygen reduction decreases the viability of the environment for aquatic organisms while the increased temperature raises the organisms' metabolic rate and need for oxygen.

3.
Temperature change — either increase or decrease — has a pronounced effect on the life-cycle regulation of aquatic organisms, and such changes may so disrupt an environment as to render it unsuitable for its natural biota even without producing a massive kill.

4.
High temperatures commonly are lethal to fish and other aquatic organisms: a fish population cannot maintain itself

continuously in an environment of elevated temperature; it must be replenished regularly from the surrounding environment.

Water Quality Standards
Let me make a few general comments about water quality criteria. These criteria for interstate waters, or portions thereof, were developed by the states in accordance with the requirements of Section 10(c) of the Federal Water Pollution Control Act, as amended. Upon adoption of criteria by a state, such criteria were submitted to the Secretary of the Interior, under a deadline of June 30, 1967, for his determination that they met the requirements of the act and of the secretary's guidelines. When accepted by the secretary, these criteria became federal as well as state standards for the water bodies in question. Thus, the determination of water quality standards is largely a matter of state determination within the guidelines established by the secretary and the act. All of the states have submitted narrative thermal standards. All of the states but two have submitted numerical temperature criteria.

By law, the Secretary of the Interior has been directed to protect and enhance the quality of the nation's waters. It is the department's position that this goal can be reached only by requiring that waste heat discharges, like other forms of pollution, be treated and controlled to such a degree that their effect will not degrade existing natural conditions or adversely affect present and potential water uses.

The present federal guidelines, prepared by the National Technical Advisory Committee, were established in April 1968 and were reviewed for applicability in April 1969 as part of a continuing effort to keep water quality criteria guidelines up to date with the most recent advances in scientific knowledge. These guidelines have been established by a panel of twenty-nine experts drawn from federal and state governments, universities, independent research laboratories, industries, and other organizations. Their opinion represents several hundred man-years of scientific experience in laboratories and in the field. They do not claim to have all the answers, but their judgment is well founded and their work will continue to provide answers to further questions.

Thermal Pollution Control

There are sources of cooling water other than natural bodies of water, for example, cooling ponds, spray ponds, and wet and dry cooling towers. Ordinarily, the most efficient and economic source is water that is piped directly from a river, lake, or ocean and then returned directly to its source. But here lies the problem. What is most efficient and economic for cooling the condenser may impair the quality of the natural water body to which the heated water is returned. The heat energy added to the water can cause problems if the speed and volume of the stream flow or tidal currents, or the method of returning the heated water, are such that the heat is not quickly dissipated.

Within a more limited scope, there may be beneficial aspects of waste heat discharges — for aquaculture, heating, extending irrigation seasons, and so forth. Clearly, then, we must seek to optimize the benefits and minimize the detriments to achieve our goal of protecting the viability of the environment while assuring continued betterment of living standards for a growing population, which includes provision for adequate electric power.

The solutions we seek can be found in today's technology. The electric power companies are accomplishing a great deal in the area of water quality protection. For example, Carolina Power and Light Company is creating a 2,250 acre lake for cooling. In Michigan, Consumers Power Company is putting cooling water back into the Tittabawassee River cleaner than it came out; further, the company is replenishing its cooling pond only during high-water-flow periods, thus aiding in the control of flooding to a certain extent. The Georgia Power Company is installing a

19 Environmental Quality and the Economics of Cooling

closed-circuit evaporative cooling system at a plant on the Altamaha River. At its new Trojan Nuclear Power Plant, Portland General Electric will use natural-draft cooling rather than once-through cooling from water drawn from the Columbia River. The Potomac Electric Power Company is designing new pumping, diluting, condensing, and discharging machinery so that the temperature rise of its discharge water will not exceed $9.2°F$. Thus the technology to deal with waste heat is clearly available and in use by a number of power companies.

Thermal pollution control adds to the cost of electricity, but the additional burden is certainly not excessive. Assuming that other charges are not affected by the type of cooling system, one can compute the percent increase in consumer costs due to the increase in production cost for thermal pollution controls from data published by the Federal Power Commission in 1966 on the consumer cost of electric power by all utilities. In Table 19.3, once-through fresh-water cooling is used as the base to which the consumer costs associated with the different cooling systems are compared.

The cost of cooling above that required for once-through fresh-water systems ranged from about 0.03 mill/kWh for once-through seawater up to a maximum of 0.2 mill/kWh for draft towers.

Table 19.3
Average U.S. Consumer Cost Increase in Electric Power with Respect to Once-through Fresh-Water Cooling (ratio)

Cooling system	Industrial	Commercial	Residential
Once-through seawater	0.34	0.16	0.14
Cooling pond	0.94	0.43	0.39
Wet mechanical-draft cooling tower	3.17	1.41	1.28
Wet natural-draft cooling tower	1.48	0.68	0.62

Source. A. G. Christianson and B. A. Tichenor, *Pacific Northwest Water Laboratory Working Paper No. 67* (Corvallis, Oregon: U.S. Department of the Interior, Federal Water Pollution Control Administration, September 1969).

These figures were derived from several independent studies of typical capital and operating costs, including consideration of fixed-charge rates, plant-load factors, and so on.

The total national cost of thermal pollution control, if all 1970 plants adopted additional cooling, would be between $45 million and $300 million.

Conclusion

Obviously we must not cast the question of balanced use of the environment in solely economic terms. Under these conditions, the value of the environment is taken at zero, and inevitably we wind up sacrificing the environment for the benefit of development. Man has existed for some two million years on the basis of his use of the environment without reckoning its true cost. Our future depends on a more reasonable assessment of environmental value.

20

Impact of Waste Heat on Aquatic Ecology Clarence A. Carlson, Jr.

In 1965, the Conservation Yearbook of the U.S. Department of the Interior stated that more water would be necessary to sustain the people of the United States for the rest of their lives than was used by all the people who previously lived on the earth.[1] Today the waters of this country are in great demand by a number of interests with needs that are frequently in conflict. The need for water for industrial, domestic, and agricultural uses is increasing, while the need for water-related outdoor recreation is also increasing. Among many water needs that might be expected to compete with recreational needs, the use of water by the power-generating industry seems likely to present a major threat. Some compromises, some changes in present policies, and some added expenses appear necessary if we are to meet satisfactorily the needs of all water-demanding interests.

The increasing human population of our country is at the root of the increasing demands for water. As crowding increases, there is a growing need for recreation, which can be supplied in part by our water resources. As one example of the importance of water-related outdoor recreation, consider sport fishing. A U.S. Fish and Wildlife Service survey in 1965 indicated that over 28 million Americans spent about $3 billion on sport fishing in that year.[2] The numbers of people who are involved in sport fishing and participate in boating, swimming, water-skiing, and other types of water-based recreation are all increasing rapidly. As our population grows, there is is also a rapidly expanding need for electricity. Analyses by Picton[3] and Trembley[4] indicate that our electrical needs are doubling every six to ten years, and there is a definite trend toward nuclear power production. Nuclear power accounted for less than 1% of our total electrical output in the

The references for Chapter 20 are on pages 360–364.

summer of 1967[5] but is expected to provide over half of our total capacity by the year 2000.[6]

Steam electric stations require water for the cooling of their condensers and have a capacity to discharge water that is warmer than the water entering them. Since power plants may interfere significantly with water quality, it is of interest to investigate the quantities of water they are expected to use. Within 30 years the power industry is expected to produce enough heat to require about a third of the average daily fresh-water runoff in the United States, if the heat were to be released to natural waters.[7] In the Northeast, over 100% of the flows in certain watersheds already pass through power stations within the watersheds during periods of low flow.[8] Stroud estimates that the 120 nuclear power plants (with combined generating capacity of 94,752 megawatts) now built or being planned would require 189,500 cubic feet per second (cfs) of water for once-through cooling.[9] Stroud has also calculated that this amount of flow represents the equivalent of 150 Connecticut Rivers, based on average annual flow in southern Vermont.

The magnitude of the expected use of our natural waters by the electric power industry is such that potential problems must not be dismissed lightly. Major debates raging in various parts of the country indicate that the possible effects of proposed power plants, especially large nuclear-fueled steam electric stations, on man and his environment are being considered carefully. Though effects vary from place to place, I would like to deal with certain basic principles concerning the significance of thermal wastes from power plants with respect to aquatic ecology, water quality, and other considerations.

In steam electric stations, heat from fossil-fuel combustion or from a nuclear-fission chain reaction in the fuel of a reactor is used to convert water to steam. The steam is used to spin a turbine that drives an electric generator. After leaving the turbine, the spent steam enters a condenser where it is converted back to water by passing over cooling tubes. The water is then pumped back to the heat source to be converted to steam, and the cycle is repeated. The condenser cooling tubes contain water that may be taken from and returned to a natural body of water. If the cooling water is not cooled before its return to a stream, lake, or estuary, the quality of the natural water can be impaired and its usefulness for other needs can be reduced. Nuclear plants waste 60% more energy (which is released to condenser cooling water as heat) than fossil-fueled plants.[7] Because larger nuclear-fueled plants are more economical, nuclear plants also tend to be several times larger in electrical output than fossil-fuel plants.[10] Amounts of thermal wastes per plant tend, therefore, to be greater in nuclear-fueled than in fossil-fueled plants. Condenser cooling water emerging from a power plant is generally 10° to 30°F warmer than when it went in (depending on the pumping rate), and it is not uncommon for this effluent to pass directly into a stream, lake, or estuary.[7]

After its discharge, the heated effluent water will float and spread in a plume on the surface of the receiving water before being carried off by prevailing surface currents. Subsequent heat dispersal depends on such factors as current speed, turbulence of the receiving waters, the temperature difference between water and air, humidity, and wind speed and direction.[7]

A healthy aquatic ecosystem is a delicately balanced combination of living and nonliving components in which stability is maintained, in part, by considerable species diversity at producer, consumer, and decomposer levels. Producers (green aquatic plants) are capable of incorporating nonliving components of the ecosystem and solar energy into food materials from which they and the consumers (herbivores) feeding on them derive energy for life processes. Energy is further transferred as secondary and higher-level consumers convert foods to their own tissues. When any of these organisms dies, its tissues are reduced by decomposers (mainly bacteria and fungi) to nutrients that can again be used by green aquatic plants. Each species in an aquatic ecosystem is dependent for its survival on the abiotic and biotic components of its environment. Subtle changes in these components may result in disruption of the system and in significant changes in the organisms that can exist in it.

Temperature is one of many abiotic environmental factors that influence the organisms in an aquatic ecosystem, and communities of aquatic organisms in various bodies of water have developed in response to temperature patterns characteristic of their habitats. Heated discharges may produce significant changes in aquatic habitats in which such communities can exist. Abrahamson and Pogue estimate that heat from a 500-megawatt nuclear plant could raise the temperature of an entire river having a flow of 1000 cfs by 16°F.[11] Cairns mentions estuarine water temperatures of 73° to 99°F near a power plant in South Wales where mean temperatures normally ranged from 45° to 63°F.[12] I recently worked as a member of a group of Cornell scientists who expressed concern about the possible effects of discharging waste heat from a proposed 830-megawatt power plant to Cayuga Lake.[13] Cayuga, like the other Finger Lakes of upstate New York, is a long, narrow, deep lake with a relatively slow flushing rate. It is stratified from May to November into an upper epilimnion and a lower hypolimnion that differ so much in temperature and density that they do not mix. The lake is essentially homothermous during the rest of the year. The proposed plant was expected to pump about 9,000 gallons of water per second from hypolimnion depth and to discharge the water (warmed about 25°F) at epilimnion depth. This would result in earlier stratification each spring, an increase in the volume of the epilimnion during each period of stratification, and therefore a substantial increase in the length of the period of thermal stratification.

Pumping of hypolimnion water might also increase the nutrients available to aquatic plants in the upper, illuminated portion of the epilimnion. These changes could be expected to increase the growing season for aquatic plants in the epilimnion, to increase the period during which hypolimnion organisms would have to subsist on a diminishing oxygen supply (replenished only when the lake waters mix), and to increase the lake's capacity for biological production. If increased biological production were realized, eutrophication of the lake would be accelerated.

Clark discusses means of minimizing effects of heated discharges by regulating the rate of flow through the condensers before release to rivers (rapid condenser water flow) and lakes (slow condenser flow), but he also states that there are many waters where no strategy of discharge will assure protection of aquatic life.[7] The possible ecological significance of mechanical damage to organisms in condenser cooling water, of chemical (including radionuclide) discharges, and of currents generated by discharges from large steam electric stations should not be overlooked.

It is well known that temperature is important to aquatic organisms as a lethal, directive, and controlling factor: lethal because high or low levels can cause mortalities, directive because it influences daily and seasonal behavior, and controlling because it affects biochemical reaction rates and metabolic rates.[14] Organisms with recreational or commercial importance may be killed by abrupt changes in temperature or by temperatures outside their limits of tolerance, and immature organisms are usually most susceptible. Temperature also affects the distribution of fish and other aquatic organisms.[15] Fish have been shown to prefer certain temperature ranges,[16,17] and an alteration in temperature may significantly alter the species composition of an aquatic community. A nonlethal increase in temperature may affect the behavior of an organism to such an extent that it is unable to escape possible sources of harm or to obtain food as effectively as it normally might. Within a tolerable temperature range, increase in temperature will increase the rates of metabolism and oxygen consumption in aquatic organisms while concurrently decreasing the oxygen-dissolving capacity of water. These factors combine to render high water temperatures less compatible with fish life.

Hemoglobin of fish blood has been shown to have a reduced affinity for oxygen at elevated temperatures.[7] Increased temperature also accelerates oxygen-consuming decay processes in water and may generally stimulate plant growth in aquatic habitats. Though elevated temperatures (within limits) increase appetite and the rate of digestion in fish, more energy is used at higher temperatures for the maintenance of body functions. Fish grown at higher tolerable temperatures tend to grow faster and have shorter life-spans than those grown at lower temperatures. Fish might also be influenced by temperature effects on their food supply. Field studies have shown that benthic invertebrates decrease in number at water temperatures above 86°F.[17] Temperature ranges within which fishes reproduce are generally narrower than ranges suitable for other activities, and developing embryos and larvae have more stringent temperature requirements than older fishes. Temperature increases may prevent reproduction and normal embryonic development or may alter spawning or hatching time to the detriment of certain aquatic species. A heated discharge may create a barrier to spawning migrations or other migrations of fish. Warming of water is also likely to increase the susceptibility of fish to certain disease organisms and to metabolic poisons. Cairns discusses the tendency for replacement of diatoms with blue-green algae as an example of competitive replacement of sensitive forms by more tolerant forms in waters that have been warmed.[12] Blue-green algae are often responsible for taste and odor problems and are less important than other algae as food for aquatic consumers.

Aquatic organisms have shown an ability to acclimate to higher temperatures, but Cairns cautions against the use of results of laboratory experiments in establishing standards for protection of aquatic organisms in natural environments.[12]

I might summarize by listing the major effects of heat pollution on higher aquatic organisms, according to Cairns:

1. Death through direct effects of heat.
2. Internal functional aberrations (changes in respiration, growth).
3. Death through indirect effects of heat (reduced oxygen, disruption of food supply, decreased resistance to toxic substances).
4. Interference with spawning or other critical activities in the life cycle.
5. Competitive replacement by more tolerant species as a result of the above physiological effects.

Cole reviews some of the biological effects of thermal pollution and also considers the global consequences of man's energy use.[18] Considerable work has been done on effects of temperature and heated discharges on specific aquatic organisms.[19–21] As of 1967, however, response to temperature had been studied in less than 5% of the almost 1,900 fish species listed in the American Fisheries Society's list of fishes from the United States and Canada.[22]

McKee and Wolf conclude that temperatures near 50°F are most desirable for drinking water.[23] Increased temperatures generally stimulate growth of taste- and odor-producing organisms, reduce the survival time of agents of disease, increase the bactericidal effects of disinfectants, and increase rates of flocculation and sedimentation in water treatment. The value of water for industrial cooling purposes and irrigation decreases as water is warmed above certain optimum levels that vary with different specific situations. Although increased temperatures have been said to be beneficial to recreation by lengthening swimming seasons, they also reduce the aesthetic value of water by stimulating the decomposition of sludge, formation of sludge gas, multiplication of saprophytic bacteria and fungi, and consumption of oxygen by putrefactive processes.[23]

Many alternatives to direct discharge of heated water to natural bodies of water have been used or suggested. Cooling water may be cooled before its return to a body of water or enclosed in a "closed-circuit" cooling system. Power plants may use artificial cooling reservoirs (1,000 to 2,000 acres for a 1,000-megawatt plant) or various kinds of cooling towers.[7] Natural- or forced-draft evaporative cooling towers have the disadvantage of discharging large amounts of water into the atmosphere. Dry towers avoid this problem by transferring heat directly to the air without evaporation, but they are 2.5 to 3 times as expensive as evaporative towers.[7,24] Comparisons of various means of cooling and their costs have been made.[17,24-26] These alternatives will reduce or eliminate hazards to natural aquatic ecosystems but will also, of course, add to the construction and operating costs of steam electric stations. The question to be answered in each situation is whether the protection of water quality adequate for aquatic life and other uses is worth the added cost. Several possible ways of using waste heat from power plants have been proposed, but none has yet found practical application. These include desalting of water, heating of buildings or greenhouses, and warm-water crop irrigation. Mihursky has suggested the creation of new ecosystems to use waste heat and organic material in the culture of human food.[14] Nash describes experiments on culture of sole and plaice in thermal discharges from power plants in south Wales and Scotland.[27] Research on using thermal discharges in the cultivation of aquatic organisms in American waters is also in progress.[26]

In view of the potentiality of thermal discharges to interfere

with various water uses, and the availability of means to control
thermal discharges, it seems reasonable and necessary to develop
thermal standards to ensure the protection of the quality of our
natural waters. I am most interested in standards to ensure the
maintenance of ecological balance and have found that adequate
standards are lacking in far too many places. Excellent criteria
for the protection of aquatic life have been compiled by a
panel of experts called the National Technical Advisory Subcommittee for Fish, Other Aquatic Life, and Wildlife. Their
recommendations, published in the 1968 Report of the National
Technical Advisory Committee on Water Quality Criteria to
the Secretary of the Interior, include the following:[28]

1.
Temperatures should not be raised more than 3°F (epilimnion
of lakes and reservoirs) or 5°F (streams) during any month of
the year, and normal daily and seasonal temperature variations
should be maintained if warm-water fish populations are to be
maintained.

2.
Inland trout streams, headwaters of salmon streams, trout and
salmon lakes and reservoirs, and the hypolimnion of lakes and
reservoirs containing salmonids should not be warmed.

3.
Discharges of heated wastes to coastal or estuarine waters should
be closely managed to avoid increases of over 4°F (fall through
spring) or 1.5°F (summer) above monthly means of maximum
daily temperatures and rates of change over 1°F/hr.

Maximum temperatures compatible with the well-being of a
few well-studied fishes are also listed, and it is noted that any
further extension of such a list is impossible because necessary
information on other species is unavailable. Although these
recommendations may not be foolproof, they are based on the
best available information and should be given full consideration
in the establishment of thermal standards. Standards for heated

discharges must also be designed to protect aquatic habitats. In thermally stratified lakes, for example, significant changes in the volumes of the epilimnion and hypolimnion by a heated discharge should be limited to avoid alteration of the entire ecosystem.

In addition to standards, there is a great need for more research on the response of various aquatic organisms to temperature. At this point I believe our lack of knowledge indicates the need for legislation to require research on the ecology of individual water bodies that are to receive large heated discharges. Additional legislation is needed to require the use of necessary technological safeguards if damaging ecological changes are foreseen. The siting of power plants should be carefully studied by specialists in many fields, and provision should be made to ensure adequate ecological research before and after plant operation.

There is reason to believe that the people of this country are finally becoming aware of the need to use their natural resources in such a way that the maximum number of people will achieve maximum benefits. A recent national public opinion poll revealed that three-fourths of the total sample of the United States population were willing to pay more taxes to protect the quality of their environment.[29] In view of this willingness to part with dollars to protect the environment and the relatively low costs of available alternatives to thermal pollution, I see little reason why we should not be able to reconcile the needs for electrical power and a variety of water uses for several years to come.

References
1
U.S. Department of the Interior, *Quest for Quality,* Conservation Yearbook (Washington: U.S. Government Printing Office, 1965).
2
U.S. Department of the Interior, *1965 National Survey of Fishing*

and Hunting (Washington: U.S. Government Printing Office, 1965).

3

W. L. Picton, *Water Use in the United States, 1900–1980* (Washington: U.S. Department of Commerce, Water and Sewage Division, 1960).

4

F. J. Trembley, "Effects of cooling water from steam electric power plants on stream biota," in *Biological Problems in Water Pollution, 3rd Seminar (1962)*, Public Health Service Publication 999-WP-25 (Washington: U.S. Department of Health, Education, and Welfare, 1965).

5

J. F. Hogerton, "The arrival of nuclear power," *Scientific American, 218,* No. 2, 21–31 (1968).

6

J. E. McKee, "The impact of nuclear power on air and water resources," *Engineering and Science, 31,* No. 9, 19–22 and 31–32 (1968).

7

J. R. Clark, "Thermal pollution and aquatic life," *Scientific American, 220,* No. 3, 18–27 (1969).

8

J. A. Mihursky and R. L. Corey, "Thermal loading and the aquatic environment. An approach to understanding an estuarine ecosystem," paper presented at International Conference on Industrial Electronics/Control Instrumentation, Philadelphia, Pennsylvania, September 8–10, 1965.

9

R. H. Stroud, "Nuclear-fuel steam-electric stations," *Sport Fishing Institute Bulletin, 202* (Washington, 1969), pp. 1–2.

10

S. F. Singer, "Waste heat management," *Science, 159,* 1184 (1968).

11
D. E. Abrahamson and R. E. Pogue, "Some concerns about the environmental impact of a growing nuclear power industry. I. The discharge of radioactive and thermal wastes," *J. Minnesota Academy of Science, 36,* No. 1 (1970), in press.

12
J. Cairns, "We're in hot water," *Scientist and Citizen, 10,* No. 8, 187–198 (October 1968).

13
A. W. Eipper et al., *Thermal Pollution of Cayuga Lake by a Proposed Power Plant* (Ithaca, New York: Citizens Committee to Save Cayuga Lake, 1968).

14
J. A. Mihursky, "On possible constructive uses of thermal additions to estuaries," *BioScience, 17,* 698–702 (1967).

15
G. V. Nikolsky, *The Ecology of Fishes* (New York: Academic Press, 1963).

16
R. G. Ferguson, "The preferred temperature of fish and their midsummer distribution in temperate lakes and streams," *J. Fisheries Research Board of Canada, 15,* No. 4, 607–624 (1958).

17
Pacific Northwest Water Laboratory, *Industrial Waste Guide on Thermal Pollution,* rev. ed. (Corvallis, Oregon: U.S. Department of the Interior, Federal Water Pollution Control Administration, 1968).

18
L. C. Cole, "Thermal pollution," *BioScience, 19,* 989–992 (1969).

19
V. S. Kennedy and J. A. Mihursky, *Bibliography on the Effects of Temperature in the Aquatic Environment* (Prince Frederick, Maryland: University of Maryland Natural Resources Institute, 1967).

20
E. C. Raney and B. W. Menzel, *A Bibliography — Heated Effluents and Effects on Aquatic Life with Emphasis on Fishes* (Ithaca, New York: Cornell University, 1967).
21
C. C. Coutant, *Thermal Pollution — Biological Effects: A Review of the Literature of 1968* (Richland, Washington: Battelle Pacific Northwest Laboratory, 1969).
22
J. A. Mihursky and V. S. Kennedy, "Water temperature criteria to protect aquatic life," in *A Symposium on Water Quality Criteria to Protect Aquatic Life,* American Fisheries Society Special Publication 4 (Washington: American Fisheries Society, 1967), pp. 20–32.
23
J. E. McKee and H. W. Wolf, *Water Quality Criteria,* 2nd ed., California State Water Quality Control Board Publication 3-A (1963).
24
W. H. Steigelmann, see Chapter 23.
25
R. W. Zeller et al., *A Survey of Power Plant Cooling Facilities* (Portland, Oregon: Pollution Control Council, Pacific Northwest Area, 1969).
26
Hearings before the Subcommittee on Air and Water Pollution of the Committee on Public Works, Parts I–III, U.S. Senate, 90th Congress, 2nd session (Washington: U.S. Government Printing Office, 1968).
27
C. E. Nash, "Thermal aquaculture," *Sea Frontiers, 15,* No. 5, 268–276 (1969).
28
National Technical Advisory Committee, *Water Quality Criteria*

(Washington: U.S. Department of the Interior, Federal Water Pollution Control Administration, 1968).
29
Anonymous, "Gallup survey on conservation," *National Wildlife*, 7, No. 3, 18–19 (1969).

21

Thermal Effects — a Potential Problem in Perspective   Walter G. Belter

In an effort to place the potential problem of thermal discharges in proper perspective, let us briefly look at the source of the problem. In a broad context, the United States faces only one problem — people and more people. Today there are 50 million more Americans than there were in 1950, and by the year 2000 there will be 100 million more people. With this growth in population come the current problems of industrial production: more automobiles, more jet airplanes, more food to be produced, more waste products, and, simultaneously, increased air and water pollution. More people means that more homes must be built, lighted, and heated; more electrical appliances, air conditioners, and so on, are needed; and all the other requirements for more electric power will increase. In this regard I should note that some thoughtful persons are now advocating that national policy should be made to discourage increases in the use of electrical energy as well as all other goods and services whose production brings added insult to our environment.
C. F. Luce has stated, "If we are to preserve a habitable earth, population growth must be stopped and members of society must be willing to except fewer goods and services — a lower standard of living — as the price of protecting the environment."[1] There are times when most of us would agree that the logic of this position is quite appealing. On the other hand, from a purely practical standpoint, I do not believe that people want to return to the horse-and-buggy days. The basic physical needs of people are such that they require light, heat, and energy, that is, electrical energy, in everything they do.

The references for Chapter 21 are on pages 384–386.

The thermal aspects of producing this electrical energy represent one of the key factors involved in the proper siting and operation of steam electric-generating plants in order to assure a minimum impact on our environment. To assess the potential thermal problem, let us briefly consider our electric power needs for the not too distant future.

Electric Power Growth

The increasing demand for electric power can be attributed to a number of factors. Population growth, of course, has been important, but it is only part of the story. Electric power usage per person has been increasing at a much faster rate than the population. Industry usage has grown. Electricity has been used in many new areas such as residential air conditioning and space heating.

Total consumption of electrical energy in the United States quadrupled between 1950 and 1968, while the population increased by one-third. The per capita consumption rose in that period from 2,000 to 6,500 kilowatt hours per year. The estimated per capita consumption in 1980 is 11,500 kWh and about 25,000 kWh by the year 2000, as shown in Table 21.1.

To meet these growing requirements, the total electric-gen-

Table 21.1
U.S. Electric Utility Power Statistics Relating to Population and Consumption

	1950	1968	Estimate for 1980	Projection for 2000
Population ($\times 10^6$)	152	202	235	320
Total power generating capacity ($\times 10^6$ kW)	85	290	600	1,352
Kilowatt capacity per person	0.6	1.4	2.5	~4.2
Power consumed per person per year (kWh)	2,000	6,500	11,500	~25,000
Total consumption ($\times 10^{12}$ kWh)	0.325	1.3	2.7	~8
Nuclear power capacity (percent of total)	0	<1%	25%	~69%

21 Thermal Effects—a Potential Problem in Perspective

erating capacity in the United States is doubling about every 10 years. The present electric-generating capacity of the United States is about 300,000 megawatts (MW). By the year 1980, the total capacity will double, and by the year 2000, the projections indicate an installed capacity in the range of 1,600,000 MW. By 1980, approximately 65% of the total generating capacity will be provided by fossil-fuel plants, 25% by nuclear plants (150,000 MW), and most of the rest by hydroelectric power. It is expected that thermal electric generation of power will approximate 92% of the total by the year 2000. At this time, nuclear and fossil-fuel plants will be providing about equal quantities of electrical energy (about 700,000 MW each).

The present trend is toward construction of large generating units of about 1,000-megawatt capacity with multiple units on a site. With this size of unit, about 540 plants would be in operation by 1980 and about 1,440 by the year 2000. If each central power station were to contain three such units, over 500 plant sites would be required by the year 2000. Although such long-range forecasts are fraught with uncertainty, it does seem reasonable to state that the number of central power station sites required to meet future power needs is sizable.

The future requirements for electrical energy and the trends in recent years with regard to plant sizes place greater emphasis on the environmental problems of plant location. The requirements with regard to air and water quality control and aesthetics place greater restraints on the selection of plant sites, especially those located near load centers. It is necessary to evaluate many environmental, engineering, economic, and sociological factors, involving many tradeoffs, in order to arrive at the proper decision for each site. Cooling-water requirements and the subsequent return of waste heat to the environment are critical factors. We are concerned about the potential environmental effects of waste heat, since these effects may result in more restrictive limitations on the long-term growth of electrical energy consumption.

Cooling-Water Needs

Along with the growth of electric power, there is a corresponding growth in the quantity of cooling water required for use in all power plants. All steam electric-generating plants release heat to the environment as an inevitable consequence of producing useful electricity. In other words, the potential waste heat problem is an energy problem and not one unique to nuclear power. In connection with attempts to assess the thermal pollution problem from these plants, it has been estimated that 200 billion gallons of water per day (bgd) will be required for cooling purposes in 1980 and 400–500 bgd in 2000. It is easy to understand why these large flows are becoming of increasing concern to water resource planners and conservationists when, according to an estimate of the Water Resources Council, the total national daily runoff of surface water in the United States is on the order of 1,200 billion gallons, which is equivalent to about 2 million cubic feet per second. The immensity of the cooling-water problem is further emphasized by the fact that the electric industry uses about 80% of all the cooling water used by all types of industry in our country, or about 40 trillion gallons per year.

Thermal Efficiencies

The quantity of waste heat that must be rejected per unit of electric power produced is directly related to the thermal efficiency of the steam generating plant, that is, the higher the efficiency, the less heat rejected. In this regard, the current light-water-reactor nuclear power plants are significantly more efficient than many of the older fossil-fuel plants still in operation. Similarly, the average efficiency of nuclear plants is approximately the same as the current average efficiency for all U.S. steam electric plants. However, the light-water nuclear power reactors currently being marketed operate at a lower efficiency and therefore reject more heat than the most modern of today's fossil-fuel plants of the same generating capacity. For this reason and because about 10%

of the heat from fossil-fuel plants is discharged directly into the atmosphere through the stack, modern fossil-fuel plants currently discharge approximately one-third less waste heat to the cooling water than do nuclear plants. The advanced reactors now under development will have heat-rejection rates comparable with the best fossil-fuel plants, and therefore the amounts of heat released to the cooling water by both types of plants will be approximately the same.

Only about 1% of the total waste heat discharge to the environment at the present time is contributed by nuclear power plants. Ten years from now, with the increasing growth of the nuclear power industry, about one-third of the total waste heat discharged to the environment by the power industry will be contributed by nuclear plants. In about 25 years, the total waste heat released from nuclear power plants will be approximately the same as for fossil-fuel plants. It is believed that this is an adequate time period for us, as engineers and scientists, to plan and develop acceptable thermal control systems, consistent with meeting our country's energy needs and providing a minimum impact on the environment.

Cooling-Water Control Technology
An important consideration in the design of every power station that will discharge heated water is the resultant temperature distribution in the receiving water. When the once-through cooling method is chosen, there are various alternatives for minimizing the impact of the heated-condenser cooling water upon a natural water body. These include the use of:
1.
Large-capacity pumps (to reduce the cooling-water discharge temperature by using more cooling water).
2.
Jet and multiport diffusers (to reduce the size of the mixing zone).

3.

Deep cold water from intake points 20–70 feet deeper than the discharge points (to use the naturally cooler water available at the deeper points and to reduce the difference between the discharge-water temperature and the natural-water body temperature at the discharge points).

4.

Dilution in long discharge canals.

Several methods of heat disposal in addition to the once-through method are available. These alternatives offer relief from thermal effects in the receiving water but involve other environmental effects and economic penalties. The alternative methods include cooling ponds and various types of cooling towers, with additional construction costs ranging from 2 to 30 dollars per kilowatt of generating capacity, as shown in Table 21.2.

In addition to the costs, there are other limitations inherent in the use of cooling towers and ponds. Although artificial lakes or cooling ponds can have very decided advantages, they can be used only where the needed land is available. For example, from 2,000 to 3,000 acres of lake surface are required for a 1 million kilowatt nuclear power plant. Moreover, substituting cooling

Table 21.2
Comparative Cost of Cooling-Water Systems for Steam Electric Plants

Type of system	Investment cost ($/kW)		
	Fossil-fuel plant	Nuclear plants	
		LMFBR[a]	LWR[b]
Once-through	2–3	2–3	3–5
Cooling ponds[c]	4–6	4–6	6–9
Wet cooling towers			
Mechanical draft	5–8	5–8	8–11
Natural draft	6–9	6–9	9–13
Dry cooling tower	17–21	17–21	25–32

[a] Liquid-metal-cooled fast breeder reactor.
[b] Light-water-cooled reactor.
[c] For 1,200–2,000-MW generating capacity.

towers for once-through cooling systems may not provide a completely satisfactory solution in all cases. A cooling tower adds large amounts of water to the atmosphere in the vicinity of a power plant, and under certain atmospheric and temperature conditions this could result in fog, ice formation on roads and powerlines, reduction in visibility, and even the formation of snow. In cooling towers, chemicals are added to inhibit biological growth, corrosion, and deposition of dissolved salts. These chemicals, transported from cooling towers by the airborne spray and by the waterborne blowdown, can have damaging effects. Also, cooling towers may pose aesthetic problems in some circumstances. Hyperbolic natural-draft cooling towers for large power plants can be over 300 feet in diameter at the base and over 400 feet high. (See Figure 17.5, page 313.)

In evaluating the use of cooling towers, account must also be taken of their demands for water. By 1990, generating units of up to 2,000 MW will be common. Water losses from cooling towers for a 2,000-MW station would be approximately 14,000 gallons per minute when the plant is operated at full capacity. Or stated another way, the quantity of makeup water for a 2,000-MW nuclear plant using a cooling tower is 20–40 million gallons per day, enough to irrigate 5,000–10,000 acres of land. Such factors must be considered when establishing site criteria for the large steam electric-generating stations of the future.

In mentioning such limitations it is important to keep them in perspective. Artificial lakes, cooling ponds, and cooling towers have decided advantages that recommend their use in certain situations. But it may not be wise to exchange the problems involved in rejecting waste heat to rivers, lakes, and oceans for the problems involved in rejecting waste heat directly into the atmosphere in the vicinity of the power plant. Certainly many of the anticipated problems involved in the use of huge cooling towers will be amenable to solution, but we must recognize that the solution will probably involve additional economic penalties.

Careful tradeoff studies will be needed to evaluate environmental, aesthetic, economic, and other factors.

AEC Program on Thermal Effects
The Atomic Energy Commission has often been accused of not being concerned over the potential thermal effects of heated water discharges from nuclear power plants. This is not true. The commission's position has been that the Atomic Energy Act does not confer regulatory authority to the AEC with respect to thermal effects, as it does with respect to radiological effects. This position was upheld in 1969 by the Justice Department. Even though the AEC has no legal authority in this area, it does take positive measures to control the potential effects of heated water discharges. For example, there is a cooperative agreement with the Department of the Interior under which the department reviews each application to build a nuclear power plant. Its recommendations on thermal effects are sent to the applicant, and the AEC urges the applicant to cooperate with appropriate state and federal agencies. Also, in 1969 the AEC testified in support of legislation that would provide for state certification to the AEC of utility compliance with state water quality standards. Such a certification would be required before a license was issued for the proposed nuclear power plant.

In addition to the regulatory activities noted, the AEC supports a considerable research and development program on thermal effects. During fiscal year 1969, approximately $900,000 was spent in the AEC program on the study of thermal effects, representing a substantial proportion of the federal effort in this area.

Participation in Joint Columbia River Study In early 1968, the AEC joined with the Department of the Interior and two of its agencies, the Federal Water Quality Administration (FWQA) and the Bureau of Commercial Fisheries (BCF), in undertaking a two-year study of thermal effects in the Columbia River. The

21 Thermal Effects—a Potential Problem in Perspective

major objective of the study as announced by the Department of the Interior was "to provide a scientific basis for determining permissible variations in stream temperatures above natural levels." In this regard, the physical and biological aspects of the Columbia River water temperatures and the effects of heat in the aquatic environment have been studied at AEC's Hanford (Washington) installation since 1946. Emphasis in the early years was on the gross aspects. The effects of elevated temperature, particularly on the Columbia River fish, were always studied along with the effects of radioactivity and chemical constituents in the effluent. However, in recent years the effects of temperature were singled out for more specific and detailed studies.

About 30 species of fresh-water fish have been identified in the Hanford reach of the Columbia River. They all are of interest, but AEC-sponsored research has emphasized the salmonid fish, especially the Pacific salmon and the rainbow trout, because of their economic importance to both commercial and sport fishing. The studies have been concerned with the effects of elevated temperature on salmon and trout as eggs, developing young, transient seaward-migrant fingerlings, transient upstream-migrant adults (returning to spawn upstream of Hanford), and as adults spawning in the Hanford reach of the river. The studies have been concerned with the direct effects of temperature on fish, the indirect effects on such aspects as the incidence of fish disease, and the combined effects of heat and other environmental stresses such as radioactivity, chemicals, and predators.

Specific AEC research that is being carried out as part of the joint AEC–Department of Interior study on the Columbia River is summarized in the following list of projects: 1. Effects of fluctuating temperatures on survival and growth of juvenile Chinook salmon. 2. Effects of temperature on the occurrence and incidence of fish diseases. 3. Relationship of thermal discharge to food and feeding of juvenile Chinook salmon. 4. Hanford Chinook salmon population studies. 5. Temperature tolerances of adult salmon. 6. Nitrogen gas bubble disease in young salmonid fish. 7. Sonic tracking of adult salmonid fish to investigate the movement of fish in the Hanford reach of the river, particularly near the reactor effluent plumes. 8. Effect of thermal shock from effluent discharge on young Chinook salmon. 9. Performance of young salmon that have been exposed to thermal shock. 10. Effects of temperature on the ability to escape predators. 11. Survival times at lethal temperatures.

Predictive Modeling of Waste Heat Discharges The AEC's Pacific Northwest Laboratory, operated by Battelle Memorial Institute, has completed a mathematical simulation of temperatures in the Illinois and Deerfield rivers below the Dresden and Yankee nuclear power plants in Illinois and Massachusetts, respectively, and also in the Upper Mississippi River Basin. This research indicates that large streams can accept and then reject considerably greater quantities of heat without exceeding water quality standards than would be indicated by routine calculations using average river flows and plant cooling requirements. This mathematical simulation technique is now being used to determine the impact of power growth, as predicted by the Federal Power Commission, on major river systems.

Temperature Prediction in Estuaries and Coastal Zones During the past several years the AEC has supported research and development work at the Chesapeake Bay Institute of the The Johns Hopkins University, directed toward predicting the physical processes of movement and diffusion in tidal areas such as estuaries. Hydraulic models and theoretical studies have been extended to develop improved methods for predicting the distribution of excess temperature resulting from the discharge of a heated effluent. Field studies to determine the temperature patterns in the receiving waters resulting from the operation of a number of large power plant installations are being planned. In these studies it is hoped that the hydraulic-model and theoretical studies that have been carried out will be validated.

Field efforts in 1969 concentrated on the Morgantown site (a fossil-fuel plant) on the Potomac estuary. It is planned that the procedures developed from studies on the Potomac estuary will be applied to predict the probable distribution of excess temperature from the proposed Calvert Cliffs Nuclear Power Plant on Chesapeake Bay. The ultimate goal will be the development of a prediction for the capacity of the entire Chesapeake Bay to assimilate safely heat releases from large steam electric power plants.

Basic Stream Data on Thermal Effects   Research by the U.S. Geological Survey (USGS), financed jointly by the AEC and the USGS, is directed toward developing more precise information about the transfer of heat between river water and air, specifically, the relationships among the eddy-transfer coefficients for heat, momentum, and vapor. Temperature surveys at eight sites having a wide variation in hydrologic characteristics are under way. Water and air temperature, humidity, wind speed, radiation, and water discharge are measured both upstream and downstream from the heat source.

The data obtained should permit a refinement in the estimate of heat lost from water to air by conduction and convection. This cannot yet be measured or computed directly but has been assumed to be proportional to heat loss by evaporation and is estimated by a ratio method. Although this has been found satisfactory for lakes, there is uncertainty about the applicability of the ratio to turbulent streamflow.

Ecological Study of South Biscayne Bay   A large complex of electrical power plants is being developed on Turkey Point in South Biscayne Bay, Florida. Two conventional power plants are now completed, and two nuclear power plants are under construction.

There is considerable concern about what the effects of waste heat may be upon the fish and other animal and plant life that now inhabit Biscayne Bay. The problem is augmented by the normally high temperature of the shallow estuary and its relatively slow flushing rate.

In May 1968, the University of Miami initiated studies of the effect of heat additions upon the ecology of South Biscayne Bay. Studies in progress include salinity and temperature mapping of the estuarine bay and studies of the relationship of water temperatures and other environmental factors to the composition, distribution, and welfare of benthic animals, benthic algae, and planktonic organisms.

Thermal Studies Extensive thermal effects studies are also being carried out by the nation's electric utilities to assure proper plant control of thermal effluents. By the end of 1969, more than 300 studies had been reported — this includes 153 studies completed, another 113 studies under way, and 41 proposed. Typical thermal effects programs for steam electric power plants being built or planned involve tidal, lake, or river current determinations; predictions of the temperature distributions due to discharge of heated waters; preoperational temperature measurements, using a variety of techniques; and ecological studies, one or two years in duration, to obtain baseline data in the vicinity of the plant site. Many operating plants are also performing postoperational ecological and temperature measurement studies to determine what changes are occurring or have occurred in the aquatic environment. These studies have resulted in changes in plant designs and, where necessary, the installation of cooling towers, cooling ponds, spray ponds, and even separate cooling lakes.

Expanded Federal Program in Thermal Effects
During 1969 I have served as chairman of a task group on the

Effects and Control of Heated Water Discharges under the auspices of the Committee on Water Resources Research of the President's Office of Science and Technology. The task group is composed of representatives of federal agencies involved in planning and supporting research and development to assure that thermal control technology is available to meet the country's needs on a timely basis. The following five areas of research require study in order to achieve an effective program in this area:
1.
Transport and behavior of heat in water.
2.
Biological effects.
3.
Treatment processes.
4.
Nontreatment solutions.
5.
Plant site selection.

Transport and Behavior of Heat in Water Various mathematical models have been developed for predicting temperature distribution and thermal behavior in quiescent waters, flowing streams, estuaries, and oceans. Improved techniques for predicting temperature behavior and for obtaining a better understanding of heated water movement are needed for early plant site evaluation and for determination of the type of effluent control system required. Data are required on the time needed to reach equilibrium conditions when heated water is discharged to the water surface, and also for mixing when heated water is diffused into a water body. Studies are needed of means for optimizing the use of mixing zones. Information is needed on the effect of atmospheric conditions on evaporative and convective heat losses for large water bodies in order to assess and improve presently available theoretical models. Procedures are needed for predicting the distribution of heated water in stratified reservoirs, including determination of the interfacial friction that occurs between a warm-water wedge and a lower layer of cold water. It is encouraging that research in several of these areas has been initiated through AEC and FWQA programs.

Biological Effects   A major problem in planning for control of the effects of heated water discharges on water quality is the lack of information on the actual temperature requirements of aquatic organisms. Each biological organism has a temperature optimum at which it reproduces and grows best. Quantitative effects on important sport and commercial fish and on fish-food organisms of temperatures above their optima are not known in sufficient detail. Without such basic knowledge, it is not possible to develop realistic water quality standards that will permit optimum use of the available water resources.

Therefore, a need exists to determine more precisely the thermal ranges required for each important species of fish and other aquatic organisms, as well as the "biological" cost of increasing temperatures at increments beyond the optimum range. This would permit selection of the temperature criteria necessary for preserving desired species. Information required includes the maximum temperature at which normal growth and reproduction can occur; determination as to whether cooler temperatures are needed for reproduction; and the magnitude of allowable temperature fluctuations. The role of temperature on fish disease should be studied further. Additional research is needed on food-chain organisms, competition, and predators, as well as the effect of heat on algal and bacterial growth. Data are needed on the avoidance reactions of fish, including field measurements as to what temperatures will be avoided and by what species of fish. Such information should indicate how much of a river or reservoir area is preempted by the heated discharges from a particular plant from use by various species of aquatic life. Quantitative biological data are being obtained on the preceding factors in the continuing Columbia River thermal effects study. Similar information is required for warm-water fish in the eastern portion of the country.

Information on the lethal temperatures for fish is known sufficiently well to define maximum allowable temperatures in the absence of other interacting factors. However, there are few quantitative data on effects of raised, but sublethal, temperatures on fish and other aquatic organisms. An increase in temperature to levels that are sublethal but still above the optimum for reproduction and growth of important aquatic organisms could result in deterioration in the species composition of an aquatic community almost as effectively as temperatures above the lethal level. The need, therefore, exists for large-scale, long-term, controlled experiments to determine the sublethal tolerable limits for entire aquatic communities. Until experiments of this type are carried out to provide factual data on the biological cost of small increases in temperature, there will be continued speculation about the importance of subtle effects, and consequently there will be controversy about the choice of standards.

Many of the needed experiments cannot be carried out in laboratory troughs or aquariums because such facilities cannot simulate in a meaningful way the natural biological communities of interest. Simulation of such communities, with rigorous control of the physical and chemical factors that influence the populations of aquatic organisms, is essential in order to provide answers to the questions concerning possible effects on the desired species and the aquatic organisms that support them. Such studies cannot be carried out on natural streams: the various habitat areas cannot be closely replicated to provide comparable control and test environments; river flows and other factors that influence aquatic populations cannot be stabilized sufficiently to eliminate their contribution to the observed result; the addition of heat (or other pollutants) in quantities sufficient to produce observable results would generally be unacceptable in the public domain.

In this connection, the FWQA in cooperation with the Tennessee Valley Authority is planning to build and operate a series of environmental test channels at the Browns Ferry Nuclear Power Plant on Wheeler Reservoir in northern Alabama. The naturalistic stream channels will provide a complete ecosystem for studying the life cycle of various warm-water fish species. Emphasis is to be placed on determining the specific qualitative and quantitative effects of heat on spawning, egg fertilization, egg and larvae development, and rates of growth.

Treatment Processes With the ever-increasing discharge of heated waters, the demand for thermal waste treatment will increase, and the need for improved and possibly new cooling methods will become more critical. Since the large thermal plants are relatively new, the limits of size, geometry, and heat loading of both mechanical- and natural-draft cooling towers are not fully determined. Further development work is also needed on the scale-up and economics of nonevaporative cooling towers. Improved design criteria for cooling ponds are required. The effectiveness of using spray ponds as a supplemental control measure for large plants should be studied. From a long-range standpoint, new and improved heat dissipation systems should be studied on a priority basis.

Various cooling methods for large-scale thermal power stations may, in themselves, create environmental contamination problems. For example, cooling tower blowdown, with its high concentration of dissolved solids and chemicals used to control organic growth in cooling-water supplies, requires treatment in order to prevent environmental pollution. Further study is also needed to quantify the effects of cooling tower operations on local meteor-

21 Thermal Effects—a Potential Problem in Perspective

ology, to ensure that another form of air pollution is not being created. Research programs in several of these areas have been started by the FWQA.

Nontreatment Solutions A most attractive method for preventing thermal degradation of water resources is through the beneficial use of the waste heat. A variety of possible uses of excess heat includes space heating, desalination of water, industrial processes, extended navigation in arctic waters, improvements in irrigation agriculture, and advances in aquaculture. Because of the thermal inefficiences of steam electric-generating plants and other industrial processes, tremendous quantities of potentially useful energy are being wasted. It has been estimated that the annual waste heat discharge that might be available for possible beneficial use will increase from the present 6×10^{15} Btu to more than 20×10^{15} Btu in 1990 (3413 Btu = 1 kWh). While the waste heat from electric power production could supply a major portion of the nation's industrial heat requirements, technical problems of power plant design and heat transport remain to be solved. In some cases, relatively low-pressure exhaust steam from thermal electric-generating plants is used in industrial processes. For example, the Dow Chemical Company plant in Midland, Michigan, is planning to use steam from a proposed power plant for process-heat requirements. However, on a national scale such use of waste heat would account for only a very small proportion of the total available supply. In some cases, it might be beneficial from a general community standpoint to reduce the efficiency of a power plant in order to supply economical heat to nearby users. This would represent a tradeoff between electric power and steam use which could be optimized at the local level.

The use of waste heat in condenser cooling-water discharge poses even greater engineering problems in possible industrial applications. Very few industrial processes can use energy of such low quality efficiently. However, agriculture is a potential user of this waste heat. Irrigation with heated water could stimulate faster seed germination and plant growth and also extend the growing season. Hothouses could be used to develop tropical or subtropical crops in the more temperate regions of the country. The solutions to problems such as soil adaptability, crop resistance to heat, and parasites would be needed before large-scale use of heated water could become common practice. A study of the possible use of waste heat for irrigation is being considered at the AEC's Hanford plant near Richland, Washington. Similarly, the Water and Electric Board in Eugene, Oregon, is considering the creation of a 2,500-acre artificial lake and use of the warm-water plant effluent for irrigation purposes.

Many areas of research exist in the potential use of condenser cooling waters for aquaculture. Marine and fresh-water organisms may be cultured and grown in channels or ponds fed with heated water. For example, it may be possible to grow commercially valuable oysters in areas where they cannot normally reproduce or survive owing to low water temperatures. The Bureau of Commercial Fisheries is proposing to join with the state of Washington in experimenting with the warm-water cultivation of oysters at Dabob Bay on the Olympic Peninsula. The bureau also plans to conduct a warm-water experiment with Coho and fall Chinook salmon in a pond near a proposed nuclear plant on the lower Columbia River. Studies are being made of the possibility of increasing lobster production in Maine with the use of waste heat. The University of Miami's Institute of Marine Science is conducting an experiment in shrimp farming at the Florida Power and Light Company's Turkey Point plant. Proposals have been made for use of warm water in developing algae, yeast, and mold cultures with agricultural and industrial wastes to produce essential protein-rich food supplements.

High priority should be given to research for the development of practical applications for effectively utilizing waste heat for industrial, agricultural, and aquacultural purposes.

Plant Site Selection As indicated earlier, several hundred new plant sites will be required by the turn of the century to meet the power needs of this country effectively. A comprehensive evaluation of plant-siting requirements should be made, using a systems analysis approach, to study the engineering, environmental, economic, and social factors involved in the identification of optimum sites on a regional basis. It is envisioned that such a program would investigate plant licensing and operating conditions, economic factors including expected changes in consumption patterns, ecological value judgments and related acceptable risks, appraisal techniques for aesthetic and recreational needs, and analytical techniques regarding thermal environmental effects. Regional studies of this type are needed, involving various federal agencies such as the AEC and the Federal Power Commission, regional power supply councils, as well as state and federal agencies administering water quality standards and conservation and wildlife resources.

Summary

The objectives of the Atomic Energy Commission's expanded thermal effects research and development program are to

1.
Develop improved water-temperature predictive techniques in streams, lakes, and estuaries (including energy budget, mixing characteristics, and velocity considerations).

2.
Encourage development of means to alleviate the thermal effects problem.

3.
Determine the effects of siting large nuclear power plants (radiological and thermal effects) on large bodies of water such as the Great Lakes and Chesapeake Bay. This study may require a systems analysis that would include consideration of engineering, environmental, economic, and social factors to provide a basis

for the identification of optimum future plant sites and would evaluate the compatability of nuclear power plant requirements and other resource utilization.

4.
Determine the biological effects of temperature and temperature changes on various water systems: fresh-water streams and lakes, bays, and other coastal environments. Emphasis will be on the effects of sublethal temperatures and on other environmental stresses that relate to the optimal growth and behavior of sport and commercial fish and supporting ecosystems.

5.
Develop beneficial uses of thermally enriched water such as agriculture, aquaculture, and possible industrial applications.

In conclusion, it is the firm conviction of the AEC that the potential thermal problem associated with a developing nuclear power industry is manageable. With good planning and continued dedicated work on the part of those in the nuclear field, the electric utilities, and those government agencies that regulate the nation's power and water systems, we can have safe, clean, and reliable nuclear power — with a minimum effect upon the environment.

References

1
C. F. Luce, remarks before the National Executives' Conference on Water Pollution Abatement, sponsored by the U.S. Department of the Interior (Washington, D.C., October 24, 1969), as reported in *Air and Water News, 3,* No. 43 (October 27, 1969), p. 5.

General reference works on the subject matter of this chapter.
2
G. T. Seaborg, "Nuclear power and the environment, a perspective," paper presented at Conference on Nuclear Power, Burlington, Vermont, September 11, 1969.

3

U.S. Atomic Energy Commission, *Nuclear Power and the Environment,* one of a series on "Understanding the Atom" (Washington: U.S. Atomic Energy Commission, Division of Technical Information, 1969).

4

G. F. Tape, "Environmental aspects of operation of central power plants," paper presented at Washington section meeting, American Nuclear Society, Gaithersburg, Maryland, December 11, 1968.

5

Sport Fishing Institute, *Thermal Pollution of Water,* Bulletin No. 191 (Washington: Sport Fishing Institute, January–February 1968).

6

U.S. Water Resources Council, *The First National Assessment of the Nation's Water Resources* (Washington, 1968). The U.S. Water Resources Council was created by the Water Resources Act of 1965. It is located at 1025 Vermont Avenue NW, Washington, D.C., 20005.

7

Clarence E. Larson, statement in *Environmental Effects of Producing Electric Power, Part I,* Hearings before the Joint Committee on Atomic Energy, 91st Congress, 1st session (Washington: U.S. Government Printing Office, 1969), pp. 214–285.

8

U.S. Federal Power Commission, Northeast Regional Advisory Committee, *Electric Power in the Northeast — A Report to the Federal Power Commission* (Washington: Federal Power Commission, December 1968).

9

J. T. Ramey, "Nuclear power, benefits and risks," paper presented at Conference on Nuclear Power and the Public, Minneapolis, Minnesota, October 11, 1969.

10
U.S. Federal Power Commission, Bureau of Power, *Problems in Disposal of Waste Heat from Steam-Electric Plants* (Washington: Federal Power Commission, 1969).

22

**Comments on the Use
and Abuse of Energy
in the American Economy**

Robert T. Jaske

The full impact of the Clean Water Restoration Act of 1966 and subsequent legislation is yet to be felt in the planning for thermal releases to the environment. Because of the long period of time between conceptual research and development in the prime mover industry, primarily electrical power generation machinery, and the actual operation of plants, the majority of problem cases under review today are products of a past era.

Increased awareness is developing within the electrical utility industry regarding the close relationship of plant thermal efficiency and policies relating to site alternatives. At present, however, the institutional framework for a fully integrated overview of environmental interactions does not exist. For this reason, the ability of investor and publicly owned utilities to meet public needs may receive serious setbacks. The meaning of conservation in the public eye has been confused by the lack of quantification of these environmental interactions among the operating entities, the equipment suppliers, and official and unofficial bodies concerned with the public interest. The strategy of industry in performance of the utility supply function primarily has involved economic optimization of fuel and equipment. In turn, fuel price is related to federal policies, subsidization, import quotas, and other indirect influences on cost. Because the smaller plants of the past had little direct thermal effect on the environment, procedures for securing approval for new facilities have been organized around mechanistic features of the system. Even in cases where determined efforts were made to enlist environmental and life-science skills in pursuit of reasonable alternatives, the existing in-

Editors' note. This paper is based on work performed under United States Atomic Energy Commission Contract No. AT(45-1)1830.
The references for Chapter 22 are on pages 392–393.

stitutional arrangements and the use of political power have militated against taking full advantage of new knowledge or even revised public opinion.

An example of the little-understood interactions among the factors involved in thermal-release planning is the question of the use of cooling towers, supported by some preservationists as a cure-all for every cooling requirement. There are three reasons why this situation needs additional study. First, once forced by circumstances to use a cooling tower, a utility must optimize the resulting plant at higher heat rates (lower efficiencies) in order to balance capital and operating costs. Low fuel prices also permit operation below thermodynamic optimums. The result is a significant increase in the total heat released to the environment, to the ultimate detriment of the public. Next, evaporation from cooling towers is significantly high to involve questions of consumptive water supply, water rights, and associated land values. Computations show that if all the projected Rankine cycle plants through the year 2000 in the coterminous United States had to use cooling towers, 12 million acre feet would be evaporated annually — equivalent to the entire annual flow of the Colorado River. (1 acre foot $= 43,560$ cubic feet $= 325,900$ gallons.) The city of Chicago is already under injunction to restrict water consumption under the Maris decision, which interpreted the 1909 treaty with Canada. Recognizing this, a major utility obtained approval of cognizant regulatory agencies for a direct discharge plant; but the opposition of other parties has led to court suits. Last, the physical appearance of huge, 500-foot-high or higher structures surmounted by cubic miles of creamy, white clouds raises serious questions of aesthetics and public safety.

Overlooked in the present subjective mood have been the significant advantages of the warming of waterways in winter, which could bring about great economic advantages to industrial communities or portions of the country where port development is affected by icing.

Studies performed for the Atomic Energy Commission using advanced simulation models indicate great public advantage in the multiple-purpose use of rivers such as the Ohio and Mississippi for direct cooling when properly augmented by ancillary cooling during critical summer months.

For the short term, predictive simulation modeling as a planning tool must be used effectively in the political system as well as in the public hearing process. Without improved public respect for the capability and quality of technical planning, the utility industry and other instrumentalities of public policy will be subject to arbitrary treatment. Careful attention by technical people to avoid inflated statements and overextrapolations can help to build public confidence in all forms of systematic management of resources.

The following general statements summarize the details of the short-term situation:

1.
Simulation modeling is fully capable of predicting and evaluating thermal effects that involve joint public and private use of thermal cooling resources.

2.
Without advance simulation, the decisions on plant size, level of treatment, treatment options, and plant location can be forced to produce built-in economic disadvantages (such as cooling towers) to the detriment of an already deteriorating rate structure.

3.
Long-range planning of industrial siting based on predictive simulation of thermal and biochemical loadings as well as other environmental effects is indispensable to all forms of municipal and industrial operation which seek to explain fully to the public the consequences of intended actions.

In the long run, the solution may lie in a complete reevaluation of the use of energy in the American economy. The concept of conspicuous consumption was first promulgated on a social basis by Veblen, but a similar view of the unlimited use of energy on a national basis may deserve consideration now (Figure 22.1). Even an early retirement of the Rankine cycle (Figure 22.2) as a prime mover for electrical generation will not solve but merely delay the problems attendant to unlimited energy consumption. With projections of total energy release to the environment from the coterminous United States exceeding 190,000 trillion Btu per year by the year 2000, with release rates in the Boston-Washington megalopolis projected to exceed 30% of the incident annual solar

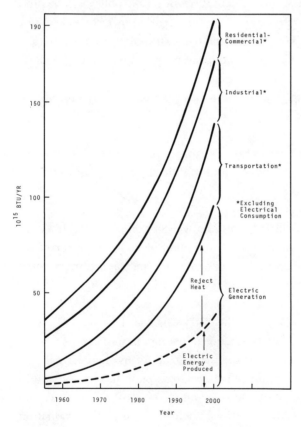

22.1
Projected total energy demand in the United States (1,000 Btu = 0.293 kilowatt-hour).

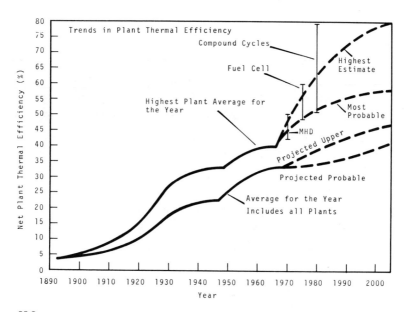

22.2
Trends in plant thermal efficiency by years.

energy by the same year, with electrical consumption of a single new building in New York soon to exceed 80,000 kilowatts, the time has come for serious examination of national energy policy on a broad front. Such an examination will soon discover that the primary motivating force in the expansion of the use of energy has been the subsidization policy of the federal government. At various historical stages in the development of the country, some form of federal inducement has been involved. As a result, the prime mover industry has been economically optimized along lines that accomplish the intended tasks set forth by government. Thus new public policy is the first step in developing a means toward a desirable long-range result.

It is suggested that such a policy should create new standards of value based on the use of thrift in resource usage rather than conspicuous consumption. The recognition of appropriate limits in the ability of the environment to absorb modification is merely a first step in this process. By use of appropriate tax incentives, such as an energy depletion allowance, such wasteful energy consumers as the disposable container, the internal-combustion automobile engine, uninsulated structures, incineration of wastes, and nonregenerative air-conditioning systems could be brought into some semblance of control. Technology is capable of eliminating the incandescent lamp in a ten-year period with attendant improvements in energy consumption.

What is lacking is incentive. It is time for national energy policy to receive thorough examination at every institutional level. We must be certain that man can maintain his well-being on earth through all the welter of obsolete inducements carried over from the Victorian notions of the Fabian Society.

References

1
R. T. Jaske, "The need for advance planning of thermal discharges," in *Selected Materials on Environmental Effects of Producing Electric Power,* Joint Committee on Atomic Energy, 91st Congress, 1st session (Washington: U.S. Government Printing Office, August 1969), pp. 523–530; also in *Nuclear News, 12,* No. 9, 65–70 (September 1969).

2
R. T. Jaske, "An independent view of the thermal effects and radioactive releases from nuclear power plants," paper presented at University of Montana Seminar on Hydrologic Problems, January 15, 1969; also available as Report BNWL-SA-2279 (Richland, Washington: Battelle Memorial Institute, January 1969).

3
William P. Lowry, "The climate of cities," *Scientific American, 217,* No. 2, 15–23 (August 1967).

4
Robert J. Nathan Associates, Inc., *Projections of the Consumption of Commodities Producible on the Public Lands of the United*

States, 1980–2000, prepared for the Public Land Law Review Commission (Washington, D.C.: May 1968). Copies of studies and reports for the Public Land Law Review Commission are available for inspection at the commission office, Room 420, 1730 K Street NW, Washington, D.C. They are also available at National Archive Federal Record Centers in other parts of the country.

23

Alternative Technologies for Discharging Waste Heat   William H. Steigelmann

The most logical way to begin a discussion of alternative technologies for achieving heat dissipation is to look at the magnitude of the problem. Industries of various types throughout the country use vast quantities of water in their various processes, and the most common use is for process cooling. The largest single use is for cycle heat rejection in electric power generating plants. In 1968, for example, 75% of the nation's total industrial cooling-water consumption — 45,000 billion gallons of water — was passed through power plant condensers.

The availability of sufficient electrical energy at low cost is essential to the economic development of any nation. Although other forms of energy are essential as well, the most flexible, most widely used, and most fundamental form of energy in our industrial base is electricity. As is well known, the historic pattern in the United States is for the installed electrical generating capacity to be doubled every ten years. The installed generating capacity of electrical utility organizations in the United States at the end of 1970 is expected to be approximately 344,000 MW, growing to 668,000 MW in 1980 and 1,261,000 MW in 1990. At present, approximately 55,000 MW (16% of the generating capability) is hydroelectric, including pumped storage, and 16,200 MW (4.7%) is from internal-combustion and gas-turbine prime movers. This means that the remainder, or about 273,000 MW, will be generated by steam-driven turbines. In 1990, the figures are expected to be:

Hydroelectric: 148,000 MW (11.7% of the total)
Internal combustion and gas turbines: 42,000 MW (3.3% of the total)
Steam-driven turbines: 1,071,000 MW (85.0% of the total).

The references for Chapter 23 are on pages 410–411.

23 Alternative Technologies for Discharging Waste Heat

Note that the percentage contribution by hydroelectric sources is decreasing with time, while the contribution from steam turbines is increasing.

Of the amounts to be generated by steam-driven turbines, nuclear-fueled plants will provide about 11,000 MW by the end of 1970, 150,000 MW by 1980, and 509,000 MW by 1990.[1] This projected growth in total electric power generating capability and the changing contribution by the various energy sources are illustrated in Figure 23.1. (Internal-combustion and gas-turbine sources are included in the fossil-fuel capability.)

Another way to look at these numbers is to consider that since existing plants will be retired at an average rate of 600–800 MW in generating capacity per year, the nation's utility organizations must build several hundred new power stations to almost quadruple the total generating capacity in the next two decades. Since adequate sites are difficult to find, most of the nearly 900,000 MW in new generating capacity will be in the form of a few

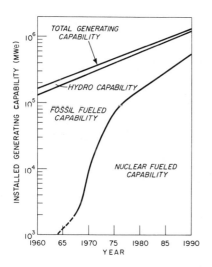

23.1
Predicted growth of electrical generating capability in the United States.

hundred large thermal power stations, each consisting of 2–4 steam turbine units that can together generate 2,000–6,000 MW at each site.

The second law of thermodynamics requires that engines of all types reject a certain amount of energy to a heat sink. In general, this heat rejection is on the order of one-half to two-thirds of the energy released in the process. Many engines, such as those in automobiles and airplanes, reject heat directly into the atmosphere, and most internal-combustion and gas-turbine power plants also do this. Since power plants using steam-driven turbines usually have a rather large generating capability, the most economic way to achieve cycle heat rejection has traditionally been to circulate water from a river, lake, or ocean through the steam condenser, which for reasons of economy is mounted directly beneath the turbine. The heat transferred to these various bodies of water is in turn transferred to the atmosphere and to outer space by radiation.

As power plants have become larger, there has developed a proper concern regarding the possible adverse affects of heated water on the aquatic environment. This, plus a growing need to utilize plant sites located at great distances from large bodies of water or large rivers, has prompted the adoption of other cooling methods: those that transfer heat more or less directly into the atmosphere. In addition to resulting in a larger capital cost for the plant, other methods of heat rejection will also require larger operating costs if the cycle efficiency is degraded because the temperature level of the cooling fluid is higher, resulting in a greater turbine exhaust pressure. A significant economic advantage results if the plant is designed to operate with a maximum exhaust pressure of 1.5 inches of mercury. This is obtainable if the condenser cooling water is brought into the system at less than about 70°F. It should also be remembered that the rate of heat transfer increases when temperature differences are increased; this works

23 Alternative Technologies for Discharging Waste Heat

against obtaining the most efficient heat cycle, and, in the case of once-through cooling, increasing the temperature of the cooling water discharged from the plant also may be damaging to the aquatic environment. Increased temperature differences do result in capital cost savings in the other methods used to achieve cycle heat rejection, however, and the optimum design is obtained by trading off decreased efficiency and higher operation costs against lower capital costs.

Public concern about the environmental effects of all types of large steam electric-generating stations has been growing during the past few years. At the same time, suitable sites for utility power stations are becoming more difficult to obtain because of the demands of industrial and general urban development. Successful planning for future generating facilities must consider the types and sizes of generating equipment that will be utilized, as well as finding a site for the plant that the public will accept and on which the plant can be constructed at as low a cost as possible while meeting all regulatory requirements.

The two most significant developments in the utility industry in the past 20 years have been, first, development of nuclear-fueled power plants, since these have exerted new competitive pressures on the fossil-fuel industry, and second, increases in the generating capability of both the power station itself and the individual units in the station. "Economies of scale" have accelerated the trend toward unit sizes greater than 1,000 MW. It is expected that this trend will continue but at a slower rate, and that the maximum size unit may increase to about 3,000 MW during the next 20 years. The size of the station will undoubtedly be dictated by both economics and possible site problems, including delivery of large equipment items and, of course, concern regarding possible adverse effects of the power station on the environment.

There is a significant difference between the amounts of heat that fossil-fueled and nuclear-fueled thermal power stations reject into the environment. This difference in cooling requirements results directly from the difference in heat rate (or thermal efficiency). This results in turn from the fact that the economically optimum cycle for the water-cooled power reactor plants requires the use of steam that is saturated or superheated only to a small extent, while fossil-fueled plants use steam at supercritical conditions. Pertinent data for the two types of plants are presented in Table 23.1. Note that the heat to be rejected from the condensers and other heat exchangers in plants that utilize water-cooled reactors is about 180% of the amount from modern fossil-fueled plants having the same electrical power output.

The figures presented in Table 23.1 pertain only to the large plants of both types that will enter service during the 1965–1980 time period. The average heat rate of all presently operating fossil-fuel plants is essentially equal to the heat rate of the nuclear plants. Similarly, it is expected that the average thermal efficiency of both fossil-fueled and nuclear-fueled plants that are placed into service during the 1980s will lie in the 40%–45% range. In the case of the nuclear-fueled plants, the efficiency improvement will result from use of reactors that are inherently capable of producing high-quality steam conditions, such as the liquid-metal-

Table 23.1
Thermal Power Plant Heat-Rejection Data

	Nuclear-fueled	Fossil-fueled
Plant net generating capability	1000 MW	1000 MW
Plant thermal efficiency	$\sim$32.5%	$\sim$40%
Plant heat rate	10,500 Btu/kWh	8,600 Btu/kWh
Total heat losses	7.1×10^9 Btu/h	5.2×10^9 Btu/h
Heat discharged directly to atmosphere[a]	$\sim 10^8$ Btu/h	$\sim 1.3 \times 10^9$ Btu/h
Heat discharged from condensers and auxiliary heat exchangers	7.0×10^9 Btu/h	3.9×10^9 Btu/h
Cooling water flow rate for 15°F temperature rise	$\sim$930,000 gpm (56 gal/kWh)	$\sim$520,000 gpm (31 gal/kWh)

[a] Heat losses through insulation around piping and equipment and from stack in fossil-fueled unit.

cooled fast breeder reactors, gas-cooled types, and perhaps the molten salt reactor.

Heat Rejection to the Environment

Heat is transferred from a body of water to the atmosphere by conduction, convection, radiation, and evaporation. The relative importance of these depends on atmospheric conditions and the solar heat load. Calculations by Shade and Smith,[2] in a discussion of a paper by Löf and Ward,[3] indicate that, on an annual average basis, approximately 50% of the heat imposed on a natural body of water is rejected by evaporation. During the winter months when the air temperature is low and the sun is in the other hemisphere, two-thirds of the heat is rejected by convection and radiation and one-third by evaporation. On an average summer day when the air is warm and the sun is high, more than two-thirds of the heat transfer is by evaporation.

When once-through cooling is employed for cycle heat rejection, in which water from a river, bay, large lake, or ocean is circulated through the steam condenser, there are some methods that can be employed to minimize the effect of the heated discharge waters on the environment. These include:

1.
Dilution of the heated stream by mixing it with a large quantity of ambient water prior to discharge. This requires an additional pumping system and a place to do the mixing (frequently the discharge canal).

2.
Use of a large volume flow of cooling water, thereby achieving a smaller temperature rise. This requires a larger condenser and larger pumps and water conduits.

3.
Discharge in a fast jet, to promote rapid mixing of the heated discharge with the receiving water, or use of a large number of discharge points. These are of limited use; unless the temperature

rise is small and movement of the receiving body fairly rapid, excessive heating of the receiving waters may still occur.

4.
Withdrawal of water from locations deep enough that its natural temperature is cooler than the surface water temperature by about the same amount as the temperature rise through the condenser. This method can be used only when cool deep water is available at the site and when the receiving body is sufficiently large that the discharged water loses its heat to the atmosphere, rather than heating the bulk of the water body. This method may cause eutrophication to be accelerated and the body of water prematurely to turn into a marsh.

When once-through cooling cannot be employed, either because of the absence of sufficient cooling water or because of adverse effects on the natural aquatic environment, other methods of cycle heat rejection must be considered, such as:

1.
Cooling pond — This features the controlled recirculation of the cooling water in a natural or man-made pond or lake, with natural heat transfer to the atmosphere.

2.
Spray pond — The heated cooling water is sprayed into the air to promote heat transfer and cooling, caught in a reservoir, and recirculated to the condenser.

3.
Combined spray and cooling pond — The addition of a spray field to augment the natural heat transfer mechanisms makes it possible to utilize a smaller pond than would otherwise be required.

4.
Natural- or forced-draft evaporative ("wet") cooling towers — These devices are essentially a spray-pond-in-a-building. The "building" is present to channel the water and air flows and to prevent excessive loss of water from the system by wind-carry of droplets, as opposed to the useful loss by evaporation. Instead of

spraying water upward, it is allowed to fall downward over a "fill" of wood or plastic partitions that serve to break up and delay the falling drops, producing more efficient heat transfer. A natural draft tower is a large chimney, about 400 feet in diameter at the base and about 400 feet high for a 1,000-MW fossil-fueled plant. A forced-draft tower utilizes fans to achieve air flow and therefore has a higher operating cost than does a natural-draft tower, although the initial capital cost is lower.

5.

Natural- or forced-draft dry cooling towers — These devices rely on the transfer of heat by conduction and convection, with no contribution by evaporation. The circulating coolant is sealed into the system.

The cooling cycle flow diagrams for the once-through river cooling system, the evaporative (wet) cooling tower, and the dry cooling tower are shown in Figure 23.2. The diagram for the once-through system would apply equally well to the cooling pond system except that in the pond there is no upstream or downstream. The cooling or spray pond, or the cooling tower, can also be used in an "open system," in addition to the "closed system" application described earlier. In many instances a plant will be located on a large body of water or a river, and the concern is with effects of the heated discharge on the aquatic life. In such cases, the pond or tower can be utilized to "precool" some or all of the condenser cooling water before it is discharged back into the river, lake, or ocean. The system can be operated "open cycle" at some times of the year and "closed cycle" at others.

Table 23.2 presents a summary of typical data on the various methods that can be utilized for heat rejection. The basic criteria for choosing among alternative methods of heat dissipation if the preferred once-through system cannot be used is evaluated cost. This decision involves in a major way the more fundamental question of where to site the plant in order to deliver a given amount of electrical energy at a load center. It is necessary to find

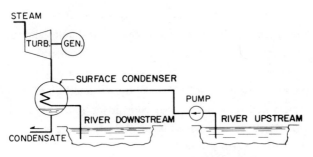

A. ONCE-THROUGH COOLING CYCLE

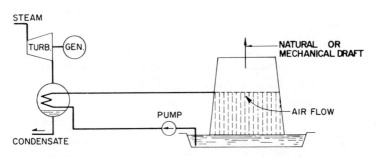

B. EVAPORATIVE ("WET") COOLING TOWER CYCLE

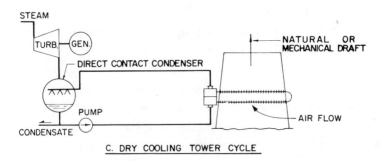

C. DRY COOLING TOWER CYCLE

23.2
Cooling cycle flow diagrams.

Table 23.2
Heat-Rejection System Data Summary (1,000 MW unit)

Type of system	Nuclear-fueled[a]	Fossil-fueled
Once-through		
Turbine back pressure (in. Hg)[b]	1.2–1.8	1.2–1.8
Land area (acres)	<1	<1
Water requirement (gpm)[c]	930,000	520,000
Investment cost ($/kW)	2–6	2–5
Cooling pond[d]		
Turbine back pressure (in. Hg)	1.5–2.0	1.5–2.0
Total area (acres)	1,500–3,000	1,000–2,000
Water requirement (gpm)	15,000	9,000
Investment cost ($/kW)	6–12+	4–10+
Natural-draft evaporative cooling tower		
Turbine back pressure (in. Hg)	1.5–3.0	1.5–3.0
Land area (acres)	~7	~3.5
Water requirement (gpm)	13,000	9,000
Investment cost ($/kW)	9–14	6–10
Forced-draft evaporative cooling tower		
Turbine back pressure (in. Hg)	1.5–3.0	1.5–3.0
Land area (acres)	~5	~2.5
Water requirement (gpm)	13,000	9,000
Investment cost ($/kW)	8–11	5–8
Forced-draft dry cooling tower		
Turbine back pressure (in. Hg)	4–8	4–8
Land area (acres)	~8–15	~4–7
Water requirement (gpm)	~0	~0
Investment cost ($/kW)	18–30	14–25

[a] Pressurized water or boiling water reactor.
[b] The units of pressure are inches of mercury (Hg).
[c] The water requirement is expressed in gallons per minute (gpm).
[d] Data are strongly dependent upon site location, topography, hydrology, and cost of land.

the optimum situation, considering such factors as the cost of delivering fuel to the plant, the cost of transmission lines, the environmental criteria that must be met, the investment cost of the heat dissipation system, the terrain of the site and cost of land, the time required to obtain approval of the site, and other less important factors. The evaluated cost must also include operation and maintenance of the heat-dissipating system and consider the effect of higher condenser cooling-water temperature on plant output and operating costs. It is usually necessary to consider the frequency with which ambient temperatures exceed certain levels to decide what is the optimum design condition. Since most of the factors are highly dependent upon the specific location of the plant, a heat dissipation method that may be economical in one instance may not be in another.

Another environmental consideration is the acceptability of a plume of vapor in the vicinity of the plant. Since the water vapor produced by evaporation is concentrated at an evaporative cooling tower or spray pond, rather than being distributed over a large area as in a once-through system or cooling pond, a visible plume of moisture will be present. Not only can this plume be aesthetically displeasing, but it can create problems by increasing the frequency and severity of fogs, rainfall, and high-humidity conditions. This can create a hazard by decreasing visibility on nearby highways, wetting of the highway surface, and icing in the winter months. For the same reasons, the plume would be objectionable if there is an airport nearby. A heat-dissipating pond or reservoir also can be expected to produce a slight increase in the frequency of local fogs in its immediate vicinity, but this is probably not a significant factor unless the reservoir is closer than a mile to the area where fog may be undesirable. Problems from plumes are more acute for forced-draft towers than for natural-draft ones, since the plume is released far above the ground.

A third environmental consideration with most heat-rejection systems is the attendant problem of chemical pollution. Chemicals such as chlorine are commonly injected into the circulating water to control the development of biological growths and marine species. Other chemicals are sometimes added to protect the condenser tubes. With evaporative cooling towers, chemicals are added to the system to protect the fill from deterioration. Because of the continuous evaporation, solids in the makeup water also accumulate and must be removed by the periodic discharge of some of the circulating cooling water to a nearby body of water. Another chemical pollution problem can occur when evaporative cooling towers are used with fossil-fuel plants because of the formation of acidic compounds when the vapor plume mixes with the plume of gases emanating from the stack. It might be noted that all of these chemical pollution problems are avoided when the dry cooling tower is used for heat rejection.

Although the dry cooling tower is presently not competitive with other methods and systems for accomplishing cycle heat rejection and therefore is not used unless water for cooling purposes is unavailable, considerable interest is being shown in the system by virtually all electric utility organizations. This interest results from a combination of at least two factors: First, the appeal of a system that does not raise thermal pollution issues, does not require water makeup, and does not create a moisture-laden plume; this can mean that a site otherwise unacceptable can be used, that many disputes with the public and regulatory authorities at hearings can be avoided, and that expensive programs to monitor the aquatic environment can be avoided. Second, since large-scale dry cooling systems for power plants have not yet been built, there is the strong possibility that technological improvements will lead to significantly lower costs.

Figure 23.3 shows the results of an internally funded study being conducted at The Franklin Institute Research Laboratories relating to improving dry cooling tower performance. The figure shows that an improvement in the overall heat transfer coefficient from 5.7 to 8.0 Btu/h-ft^2-°F will result in a cost reduction of about 25%. It also illustrates the strong dependence of cost on design temperature; the same 25% savings being achieved by a 10°F temperature reduction.

Uses for Rejected Heat
I would also like to mention the application of binary-cycle plants.

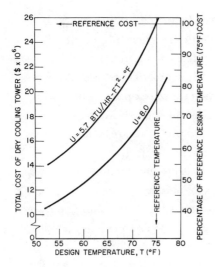

23.3
Dry cooling tower cost as a function of design temperature and overall heat transfer coefficient.

These offer the potential of a power conversion system that makes possible more economical dry cooling towers. As part of a research program sponsored by the Office of Coal Research, U.S. Department of the Interior, The Franklin Institute Research Laboratories has made an engineering analysis of the application of an ammonia turbine bottoming cycle, in connection with conventional steam turbines and dry cooling towers, as a means of reducing both cooling-water requirements and the cost of generating electricity in coal-fired power plants.[4] The use of a dry cooling tower for condensation of the ammonia vapor could well provide the most economical system for mine-mouth sites in arid, coal-bearing regions. The binary-cycle power conversion system, possibly using a less toxic fluid than ammonia, is potentially attractive for wide application throughout the United States, however, since it offers a way to accomplish cycle heat rejection, which gives the least perturbation to the natural environment.

To date, dry cooling towers for condensing steam have not been widely used in this country because of their high capital cost, which results from the large and expensive components needed to handle the large volume of low-pressure steam from the turbine exhaust. However, the bulk and cost of the dry cooling tower system can be greatly reduced when an ammonia turbine is substituted for the low-pressure steam turbine; ammonia vapor has a much lower specific volume than steam at turbine exhaust conditions. For the same reason, the low-pressure turbine is significantly smaller and less expensive in the binary-cycle system.

Although the binary cycle was not fully optimized, the study comparing different mine-mouth plants showed that a 1,000-MW steam-ammonia power plant would require a smaller capital investment than a conventional, all-steam plant, when both were sited where dry cooling towers were required. Future efforts on this subject will involve evaluation of alternative fluids for the bottom component, investigation of the benefits of superheating the bottom-component fluid, and studies pertaining to the development of improved and more economical cooling towers for cycle heat rejection. The results of a study of the application of a binary-cycle concept using refrigerant gases to nuclear power plants were presented by El-Ramly and Budenholzer at the 1963 American Power Conference.[5]

There have been a number of suggestions for beneficial uses of the heat rejected from power-generating facilities. Potential uses of the heated discharge from a power plant include:

1.
Aquaculture — production of shrimp, catfish, and gamefish during the winter.
2.
Agriculture — extending the growing season for crops and preventing damage from early or late frosts.
3.
Upwelling in the oceans — mixing of nutrient-rich water from the cool depths with less rich warmer surface waters.

4.
Heating of buildings.
5.
Maintaining streets and sidewalks clear of ice and snow.
6.
Cleaning and warming the air in a city by releasing heated air at various points.

A number of studies and experimental and pilot-plant programs have been carried out pertaining to these ideas, and in many instances the results are promising.

Conclusion

I would like to conclude by quoting from an account of a talk given by Charles F. Luce, board chairman of Consolidated Edison Company of New York. The occasion of his talk was the two-day Executives' Conference on Water Pollution Abatement, called by Secretary of the Interior Hickel in October 1969. "If we are to preserve a habitable earth, population growth must be stopped and the members of society must even be willing to accept fewer goods and services — a lower standard of living — as the price of protecting the environment." The logic of this position is quite compelling, he said, "but for those of us who manage public utilities . . . the logic is irrelevant. Con Edison cannot tell New Yorkers that it is better to have fewer power plants and transmission lines than to be cooled by air conditioning." He repeated the contention that doubling energy supply every 10 years will affect the environment, with the only question being how much change is acceptable.[6]

I believe that we must accept the point that Luce is trying to make: we have to determine the extent to which we want our environment changed from the way it was a few hundred years ago. This does not mean that we should or must accept a "polluted" or "contaminated" environment, but rather that we must determine at what point we judge that a changed environment becomes

23 Alternative Technologies for Discharging Waste Heat

a polluted one, recognizing that some change is inevitable, that not all change is pollution, and that we are all going to pay for the maintenance of the environment. We must also recognize that at present there are insufficient scientific facts regarding the effects of heated discharges on the aquatic life in a body of water on which to base accurate decisions. At this stage, it is natural that experts will sharply disagree on quantitative predictions of the effect of a power plant on the environment.

In summary, these are the pertinent factors to consider:

1.
There is clearly a need for more power plants; we cannot delay their construction since the ever-increasing need for electric power is already placing a strain on today's installed generating capacity, and reserve capacity is less than it properly should be in the opinion of the Federal Power Commission.

2.
These power plants will be large — because of scarcity of sites and economies of scale — requiring the dissipation of large amounts of energy at each site.

3.
It is known that too high a temperature will kill aquatic life and adversely change our rivers, lakes, and coastal bays, but precise limits are not known. Long-term and detailed ecological studies are required. (It should be noted that many individual utilities, the Edison Electric Institute, the AEC, and the FWQA have been and are continuing to fund such studies.)

4.
The regulatory situation is in a state of flux, and in many instances the utility does not know what water quality standards will be made to apply to its project even though its plant may be in an advanced stage of construction.

5.
No one — engineer, scientist, or utility executive — wants to degrade the natural environment to a noticeable extent.

This is our dilemma: How should the regulations that are to protect our environment be written in order to accomplish their goal without delaying construction of our needed generating facilities and without requiring that excessive and unnecessary costs be incurred, which would also be a waste of our nation's resources? We are faced with a rather serious problem in attempting to solve the dilemma. The first step toward a solution is to educate the scientific community through an enlightened and rational discussion of the situation and, then, the public and those who enact our laws. I hope that the symposium on power generation and environmental change and this volume will help to advance us toward this goal.

References

1
F. Stewart Brown, in supplemental material accompanying statement of John N. Nassikas in *Environmental Effects of Producing Electric Power, Part I,* Hearings before the Joint Committee on Atomic Energy, 91st Congress, 1st session (Washington: U.S. Government Printing Office, 1969), p. 56.

2
W. R. Shade and A. F. Smith III, in discussion of "Economic considerations in thermal discharge to streams," *Engineering Aspects of Thermal Pollution, Proceedings of the National Symposium on Thermal Pollution,* August 14–16, 1968 (Nashville: Vanderbilt University Press, 1969).

3
George O. G. Löf and John C. Ward, "Economic considerations in thermal discharge to streams," *ibid.*

4
Z. M. Slusarek, *The Economic Feasibility of the Steam-Ammonia Power Cycle,* The Franklin Institute Research Laboratories Report No. PB184331, prepared for Office of Coal Research, U.S. Department of the Interior (1968).

5
Nabil El-Ramly and R. A. Budenholzer, "Binary cycle for nuclear power generation using steam and refrigerant gases," *Proceedings of the American Power Conference, 25,* 496–504 (1963).
6
C. F. Luce, remarks before the National Executives' Conference on Water Pollution Abatement, sponsored by the U.S. Department of the Interior (Washington, D.C., October 24, 1969), as reported in *Air and Water News, 3,* No. 43 (October 27, 1969), p. 5.

Glossary of Nuclear Terms and Units

The various nuclear units and terms (curie, roentgen, rad, rem, and RBE) relate to the measurement and specification of three physical quantities: the amount of a nuclide with a certain level of radioactivity, the amount of ionizing radiation at a particular location, and the radiation dose in a particular medium. Nuclide refers to a particular atomic species characterized by an atomic number Z equal to the number of protons in the nucleus and a mass number A equal to the total number of protons and neutrons in the nucleus.

Curie (Ci)

This unit is the amount of a radioactive nuclide that undergoes 3.700×10^{10} radioactive transformations per second. The unit honors Marie Curie and was originally related to the activity of a gram of radium.

At any instant, the number of curies represented by a mass W of a nuclide of atomic mass M is equal to the rate at which the total number of nuclides will undergo radioactive transformation, divided by 3.700×10^{10}. The number of curies is customarily referred to as the activity of a nuclide. It includes all competing modes of radioactive transformation:

$$\text{Activity} = \text{number of curies} = \frac{W(g)}{M(g)} \frac{6.025 \times 10^{23}}{3.700 \times 10^{10}(\text{sec}^{-1})} \lambda(\text{sec}^{-1})$$

where λ is the total decay constant, which can be thought of as the probability per unit time for radioactive transformation. It is equal to 0.693 divided by the half-life $T_{1/2}$ expressed in seconds.

The curie specifies only the number of disintegrations. It is not a measure of any effect or hazard.

One curie of uranium-238 ($T_{1/2} = 4.51 \times 10^9$ years) corresponds to 3,000 kg of uranium-238. One curie of radium-226 ($T_{1/2} =$

1.62 × 10³ years) corresponds to slightly more than 1.0 g of radium-226. One curie of strontium-90 ($T_{1/2} = 20$ years) corresponds to 5.0 mg of strontium-90.

Roentgen (R)

The roentgen is a unit of energy dissipated in dry air at standard conditions by a beam of x rays or gamma rays. It is customarily used to specify quantity of radiation. The unit commemorates Wilhelm Konrad Roentgen, the discoverer of x rays.

One roentgen is the quantity of radiation that will produce 1 electrostatic unit charge of electricity of either sign (or 2.08×10^9 ion pairs) in 1 cm³ of dry air at 0°C and 760 mm Hg. Assuming that 34 eV are required for the formation of an ion pair, 1 roentgen is equivalent to an energy dissipation in air, or dose, of 87.7 ergs/g of air.

In order to calculate the quantity of radiation dissipated in a volume of air at a certain distance from a radioactive source whose activity is known in curies, it is necessary to know the spectrum of the emitted radiation, and the absorption and scattering coefficients. As an example, it can be shown that a 1-curie point source of 1 MeV gamma rays would produce a dose of 0.5 roentgen in 1 cm³ of air at a distance of 100 cm in 1 hour.

Rad

The rad is a unit of energy dissipation, or radiation dose, in any medium. It is applicable to any type of radiation, whereas the roentgen is restricted to x rays and gamma rays. 1 rad is equal to an energy absorption of 100 ergs/g of the medium.

The medium is frequently human tissue. The dose in rads to human tissue, placed at a point where the amount of x or gamma radiation over a period of time would be 1 roentgen, and it would vary depending on the type of tissue and the spectrum of radiation. However, for soft tissue, the dose corresponding to 1 roentgen of radiation would be approximately 100 ergs/g of tissue, or 1 rad.

RBE

The dimensionless quanity RBE stands for the relative biological effectiveness of various radiations with respect to 200-keV x rays. The RBE for x, gamma, and electron radiation is approximately 1. The RBE for protons, fast neutrons, and alpha particles is, typically, 10. That is, fast neutrons can be ten times more effective in producing biological change than gamma rays.

Rem (rem)

Rem is an acronym for "roentgen equivalent man." The number of rem is equal to the number of rad times RBE. The rem refers to all types of radiation and is frequently applied to total body exposure to mixed radiations. It is the unit in which government standards for protection are expressed.

Index

Abortion, radiation-related, 86–88
Acid mine drainage, 320, 329–336
  methods and costs of control, 332–336
    table, 334
  from refuse banks, 325
  from strip mines, 329–330
  from underground mines, 330–331
Acidity, atmospheric, 297, 299, 302
  in precipitation, 299
    figure, 300
Acrolein, in smog, 278
Activation gases released from reactors, tables, 36, 115
Activation products in reactors, 25
Advisory Committee on Reactor Safeguards (ACRS), 31
Aerosols, atmospheric, effect on heat balance, 254–255, 260
Agriculture
  effect on world climate, 254–255, 260–261
  use of waste heat for, 358, 381–382, 407
Air. See Atmosphere
Air-water interface, energy transfer across
  effect of oil pollution, 254–255
  effect of thermal pollution, 375, 399
  evaporation loss, 138, 145, 399, 404
Akosombo, 152
Alabama, 380
Albedo, and heat balance, 254–255, 260–261
Aldehydes, in smog, 275, 277–278
Algae, in thermally polluted water, 356
Algeria, oil exploration in, 228–229
Alkali salts in combustion gases, 203
  figure, 203
  and MHD, 204, 217–218
Allen, Eric R., 263–288
Alpert, Seymour B., 228–245
Altamaha River, 349
American Fisheries Society, 357
Aminoil, 235
Ammonia (NH_3)
  atmospheric, 271, 297

  in smog, 275
  in stratosphere, 281
Ammonia turbine bottoming cycle, 406–407
Ankylosing spondylitis, 65
Anthracite mining
  refuse banks from, 323–326
  as fuel, 192
  photographs of, 324, 326–328
  subsidence from, 322–323
    photograph, 323
Antimony-125, 75
  bone dosage, 46–47
  maximum permissible concentration, 46
Appalachian Regional Commission, 330, 335
Aquaculture, with waste heat, 382, 407
Aquatic ecology
  need for research on, 360, 382
  and thermal pollution, 353–358
Aquatic nutrients in reservoirs, 139, 142–144
Aquatic plants in reservoirs, 144–146
  and evapotranspiration, 138, 145
Area strip mining, 318. See also Strip mining
  photograph, 319
Arthur Kill power plant, 312
Ash. See also Fly ash
  fusion temperature, 14, 181–182, 191–192
  removal of
    in MHD generation, 218
    by scrubbing, 204
  sintering temperature, 191–192, 194–195, 199
  sodium content of, 203–204
    figure, 203
Aswan dam, 138, 143
Atlantic City Electric Company, 313
Atmosphere, 246–261
  carbon dioxide content, 247–248
    thermal effects of, 251–254
  historical temperature change of, 253
  oxygen content, 261

Atmosphere (continued)
  regions of, 264-265
    polluted, 274-279
    troposphere, 266-279
  temperature and pressure of, figure, 265
  thermal balance of, 248-249
Atmospheric chemistry, 263-285
  of air pollution, 274-279
  neutralization of acid and base, 297, 299
  of nitrogen compounds, 271-274
  of sulfur compounds, 268-271
Atmospheric pollution, 274-279
  composition of, table, 275
  daily variations in, figure, 276
  gaseous reactor waste, tables, 29, 36, 115
  nitrogen oxides, 294-295, 297-300
  organic compounds, 273
  polluted precipitation, 292-299
    total excess acid and base in, figure, 300
  sulfur, 302
  sulfur dioxide, 289-299
Atomic Bomb Casualty Commission, 49
Atomic Energy Commission (AEC). See under U.S. Government
Aufwuchs, 144-146
Automobile exhaust, 274, 297
Avco-Everett Research Laboratory, 218

Background radiation, natural, 31, 95-96
  biological effects of, 130
Baltimore Gas and Electric Company, 94, 313
Bandama Lake, 152
Barium-140, in liquid reactor waste, 28
Battelle Memorial Institute, 374
  fluidized bed of, 204
BCURA Industrial Laboratories, 193-199, 218-219
  high-pressure fluidized bed boiler figure, 193
  photograph, 194
Bechtel Corporation, 233
Behnke, Wallace B., Jr., 11-19, 209
Bell, Earl J., 127
Belter, Walter G., 127-130, 365-386
Benedict, Manson, 214
Benthos, 144
  growth in heated waters, 356
Berkowitz, David A. (coeditor), 158-172
Big Rock power plant
  noble and activation gases from, table, 115
  radioactive effluent from, tables, 36
Bilharziasis, 145
Binary cycle power plants, 405-407
Biological concentration
  in food chain, 44, 47, 77
  in tissue, 32-33
Biological effects of radiation
  on birth rate, 85
  on birth ratio (male-to-female), 56-57, 69-71, 85-91
    table, 87
  disease, 49-52, 64-68
  on fetus, 50-51, 53-58, 62-72, 85-87
  genetic, 54-58, 68-71, 96-99
  from natural background, 130
  on oocytes and oogonia, 71-83
  on reproductive performance, 83-84
    table, 87
  superovulation, 83, 88
Bituminous mining waste, 323-326
Black, Sivalls, and Bryson, Inc., 219
Black River, East Fork, 169
Blue Coal Corporation, 192
Boiling water reactor (BWR). See Nuclear reactors, boiling water
Bone, radiation exposure
  from air near power plant, table, 111
  cause of cancer, 50
  from radionuclides in water, 32, 46-47
    table, 46
    discussion of table, 75-76
Boston Edison Company, 241
Boston-Washington megalopolis, energy requirements, 390

Index   417

Bottoming cycles, 405–407
Breast cancer, radiation-induced, 50
Breeder reactor. *See* Nuclear reactors
British Coal Utilization Research Association. *See* BCURA Industrial Laboratories
British Esso, 207–208
Browns Ferry nuclear power plant, 380
Bureau of Mines. *See under* U.S. Government, Department of the Interior

Cadle, Richard D., 263–288
Calcination, for sulfur oxide removal
 in fluidized beds, 206–208, 218
 in PF furnaces, 188
California Water Project, 163–164
Calvert Cliffs nuclear power plant, 94, 375
Cancer
 doubling doses for, table, 50
 discussion of, 64–66
 radiation-caused, 49–52, 64–68
 in young people, 50
 risk of, tables, 98–99
  compared with other risks, figure, 100
Carbon, earth resources of, figure, 248
Carbon-14, atmospheric, 120
 dosage from, 120
Carbon dioxide (CO_2), 120, 246–259
 atmospheric content, 120, 247–251, 283–284, 289
 released by man, 247, 250
 thermal and climatic effects of, 250–254, 284
 yearly increase of, 247–248
 in oceans, 251
 partition in environment, 250
  figure, 248
 in smog, 275
 in stratosphere and mesosphere, 283–284
 Suess effect, 120
Carbon monoxide
 in atmosphere, 273, 289
 in smog, 275

 reactions of, 278
 in stratosphere and mesosphere, 283
Carcinogens from fuel combustion, 41
Carlson, Clarence A., Jr., 351–364
Carolina Power and Light Company, 348
Castaic reservoir, 164
Catalytic desulfurization, 230–234
 figure, 232
Catalytic oxidation, 187
Cayuga Lake, 354
Central nervous system, radiation-induced tumors in, 51
CERCHAR (Centre d'Etudes et Recherches des Charbonnages de France), 200–203
Cerium-144, in reactor liquid waste, 28
Cesium-134, in reactor liquid waste, 28
Cesium-137
 maximum permissible concentration, 46
 reactor inventory of, 47, 77
 in reactor liquid waste, 28, 76–77
Chemostratification, 139
Chesapeake Bay, 94, 375, 383
Chesapeake Bay Institute, 374
Chevron, 234
Chicago, water consumption, 388
Chlorine, in cooling towers, 404
Chromosomes, radiation damage in, 55–58, 68–72, 83–91
Cities Service, 234–235
City College of New York, The, 207–208, 238–239
Claus-Chance process, 207–209
Climate, 246–261
 effect of carbon dioxide on, 250–254
 and landscape change, 254–255, 260–261
 and man, 253
 need for global study of, 257–259
 observed changes in, 252–255
Coal
 energy and chemicals from, 11, 17, 209
 new emphasis on, 213–217

Index

Coal (continued)
  oil and gas conversion from, 15, 209–213, 215
  radioactive fly ash content of, table, 108
  sulfur content of, 307
    effect on precipitators, 183
    figure, 186
    low-sulfur reserves, 14
Coal consumption
  global, 249
  in U.S., 13, 317
Coal-fired power generation
  history of, 177–182
  projected increase of, 16
Coal-fired power plants, 10–18, 177–182, 213–217
  power and chemicals from, 17, 209
  radioactivity from, compared with nuclear, 107–123
    tables, 111, 118, 120
  stack height and radiation exposure, 119–120
Coal-firing techniques
  fluidized bed, 189–195
    high-pressure, 193–202
  gas temperature, 181, 191–192, 195, 199–201
  history of, 177–182
  in molten iron, 219
  pulverized fuel (PF), 177–189
  Szikla-Rozinek, 203
Coal gasification, 205
Coal mining, environmental aspects, 317–338
  acid mine drainage, 320, 329–336
  mine fires, 323
  refuse banks, 323–326
    photographs, 324, 326–328
  strip mining, 319–322, 326–329
    photographs, 319–321
  subsidence, 322–323
    photograph, 322
Coal production, 250
  from strip mines, 317–318
  from underground mines, 323
Coal reserves, 14–15, 250
  low-sulfur, 14

Coalplex, 209–213
  figures, 210, 212
Cobalt-58, in reactor liquid waste, 28
Cobalt-60, in reactor liquid waste, 28, 75
Code of Federal Regulations, radiation standards, 44, 46–49
Coke
  from coal desulfurization, 210–211
  historical use and production, 179
Colorado River, 388
Columbia River, 349, 373, 379, 382
Combustion Engineering, Inc., 188–189, 241
Commonwealth Edison Company, 11–12, 14–15, 110, 241
Concentration of radionuclides, biological
  in food chain, 44, 47, 77
  in tissue, 32–33
Connecticut River, 163
  flow-rate, 167, 352
Connecticut Yankee power plant, 118
  noble and activation gases from, table, 115
  radioactive effluent from, tables, 36
Conowingo dam, 164, 167
Consolidated Edison Company of New York, 161, 182–186, 228–229, 312, 408
Consolidated Coal Company, 206, 208, 210, 331
Consumers Power Company, 348
Contour mining, 318–319. *See also* Strip mining
  photograph, 320
Coolant disposal, reactor, 27
Cooling
  once-through, 346, 348–349, 353, 369–370, 399–400
    alternatives to, 400–401
  of power plants, 341–350, 352, 368
    chemical pollution from, 404–405
    investment costs of systems for, tables, 370, 403
    water flow rates required, 398
Cooling ponds, 348, 358, 370, 400–404

Cooling ponds (*continued*)
  flow diagram, 402
  investment cost, tables, 370, 403
Cooling towers, 348, 358, 370–371, 400–407
  dry, advantages of, 405–407
  dry, cost vs. heat transfer in, 405
    figure, 406
  flow diagram, 402
  investment cost, tables, 370, 403
  photograph of, 313
  wet, environmental effects of, 371, 388, 404
Cooling water, 344–346, 352, 368, 394, 399–400
  tables, 345, 398, 403
  used by U.S. industry, table, 345
Cornell University, 354
Cornwall pumped storage project, 161
Corrosion products from reactors, table, 36
COWAR. *See* International Council of Scientific Unions
Cracking, to remove sulfur from oil, 239–240
Crude oil. *See* Oil
Cryogenic sampling, 266, 283–284
Curry, James A., 320
Cycle heat. *See* Waste heat

Dabob Bay, 382
Dams, 133–154. *See also* Dams, effects of
  economic, social, ecological effects of, figure, 137
  generating capacity of, 159
  hydroelectric potential of U.S., 159, 161
  and national pride, 135
  photographs, 134, 140–141
  purposes for, 135
  reservoirs and lakes compared, 135
Dams, effects of. *See also* Dams
  atmospheric, 139, 141
  biological, 141–147
    on aquatic ecosystem, 144–147
    on terrestrial ecosystem, 142–143
  economic, 148–150

  geological, 141
  on humans, 147–152
    disease, 145–146, 149
    industrialization, 149
    psychological, 147–148
    social, 147
  hydrologic, 136–139
  on transportation and communication, 147–152
  photographs, 143, 151
Davis, Billy, 319, 321
Davis, H. L., 5
Deerfield River, 374
Deforestation, effect on climate, 260–261
DeHaan, Robert L., 95
Del value for sulfur, 303
  in fuel oil, table, 309
  table, 304
Delaware River, 160, 163
Delmarva Power and Light Company, 313
Delta-Mendota canal, 164
Denmark, precipitation, 292–299
  nitrates in, figure, 298
  sulfur in, figures, 293, 296
  total excess acid and base in, figure, 300
Desulfurization of oil, 230–242
  cost of equipment for, 249
  plants for, table, 235
  at power plant, 235–242
  at refinery, 231–235
Detroit Edison Company, 188
Diagnostic use of x rays, 50–52, 69
Diamond, Earl L., 69–71, 82–93
Diseases caused by radiation, 49–51, 64–68
  genetic hazards, 55, 85
  risk of, tables, 98–99
    compared with other risks, figure, 100
  in young people, 50
Distillate, percent in oils, 231
Distillation of acid mine water, 334
Dolomite sulfur oxide removal
  in fluidized bed, 206–208, 218–219
  in oil combustion, 238

Index

Dolomite sulfur oxide removal (continued)
  in pulverized fuel furnace, 188
Dose, radiation, 54, 67
  to bone and lung from air near power plant, table, 111
  to bone from radionuclides in water, 46–47, 75–76
    table, 46
  lethal, for oocytes, 84
  threshold, 53–54, 68, 79
Doubling dose, 52–53
  discussion on use of, 64–66, 80
  table, 50
Dow Chemical Company, 381
Dresden Nuclear Power Station, 11, 110, 118
  noble and activation gases from, table, 115
  radioactive effluent, 113–115
    tables of, 36, 114, 115
  site characteristics, 110
  whole body dose rate in plume from, 115
Dry cooling towers, 405–406. See also Cooling towers
DuBridge, Lee, 11
Dunster, H. J., 103n
Dust, effect on climate, 254–255, 260
Dust-collection equipment. See Electrostatic precipitators
Dutch Shell, 240

Earth, energy balance for, figure, 256
Ecology
  aquatic, and thermal pollution, 353–358
    need for research in, 360
  riverine and lacustrine, 136–139, 145
Edington, Charles W., 63–74
Edison Electric Institute, 409
Efficiency of heat rejection cycles, 396–397
Efficiency of power generation
  with bottoming cycles, 406–407
  figure, 391
  with gas turbine or MHD topping, 217
  nuclear, 213–214, 398
  nuclear and fossil compared, 344n, 353, 368–369, 397–398
  pumped storage, 159
  relation to cooling cycle, 396
  and siting requirements, 387–388
Eisenbud, Merril, 23–43, 75–77, 80–81, 126
Electric power. See Power production
Electrodialysis for acid mine water, 334–335
Electrostatic precipitators, 13, 183–186
  cost of, figure, 186
  design curves, figure, 185
  operating temperature, 183
  performance for low-sulfur coal, 183
  photograph, 184
  size, figure, 183
Elk River power plant, radioactive effluent, tables, 36
Energy balance, earth, 255–257, 260–261
  and air-water energy transfer, 254–255
  effect of carbon dioxide, 251–254
  effect of gaseous pollutants, 248–249
  figure, 256
  need for study of, 257–259
  perturbed by man, 255–257
Energy production
  atmospheric heating from, 254–257
  cooling water requirement, 344–346, 352, 368, 394, 399–400
    table, 345, 398, 403
  dependence of society on, 7–10
  global, 249
  per capita, 256–257, 366
  large vs. small plants, 12
  U.S., 3, 12, 16, 249, 351, 366–367
    in Boston-Washington megalopolis, 390
    figures, 390, 395
    installed capacity, 394–395
    and national policy, 7–10, 391–392
    skyscraper requirement, 391
    tables 344, 366

Index   421

Energy resources
  coal or nuclear, 213–216
  global consumption, 248–251
  importance of competition among, 16
  management and public policy, 391–392
  U.S., 317
  wasteful use of, 392
Energy transfer, air-water
  evaporation loss, 138, 145, 399, 404
  and oil pollution, 254–255
  and waste heat dispersal, 375, 399
England, sulfur in precipitation, 294
  figures, 295, 296, 300
English Station power plant, photograph, 312
Environment
  man's interaction with, 166, 170–171, 253
  man's value toward, 3–5, 350, 360, 365, 408–409
Environmental alteration, program for study of, 257–259
Environmental degradation
  a polluted atmosphere, 274–279
  cost of cleanup, 8
  cost-benefit of improvements, 336–337
Environmental monitoring near reactors, 31, 33
Environmental quality
  and power policy, 10, 387–388, 391–392
  and societal conflicts, 8, 343
Environmental radioactivity, 31, 95–96
  in soil, 110
Eriksson, Erik, 289–301
Erosion
  caused by strip mining, 319–320
    photograph, 321
  related to dams, 138, 146, 152
Esso Research and Engineering, 234
Europe, atmospheric pollution
  from nitrates, 297
    figures, 298, 300

  from sulfur, 292–294
    figures, 293, 295–296, 300
Eutrophication and thermal pollution, 355–356, 400
Evans, Robley D., 79
Evaporation loss, 138, 145, 399
  from cooling ponds and towers, 404
Evaporative cooling. *See* Cooling ponds; Cooling towers
Evapotranspiration, 138, 145
Exhaust, automobile, 274, 297
  and photochemical smog, 274–276
Eye irritation from smog, 278

Fallout from weapons tests, 35, 76
Fanning, Delvin S., 261
Farmer, Robert E., Jr., 337
Fast-breeder nuclear reactor
  projected use, 15–16
  uncertainties, 214
Fay, James A., 3–6
Feather River, 163
Federal Power Commission (FPC). *See under* U.S. Government
Federal Radiation Council (FRC). *See under* U.S. Government
Federal Water Quality Administration (FWQA). *See under* U.S. Government, Department of the Interior
Fetal death, radiation-related, 86–88
Fetal germ cells, radiation effects, 56, 82–83
Fetus, radiation effects, 66, 68–69
Filtration of reactor coolant, 27
Finger Lakes, 354
Fish aquaculture in heated water, 358
  proposed study of, 382
Fish migration, effect of dams and pumped storage on, 146–147, 168
Fish populations
  effect of flood control on, 146
  effect of thermal pollution on, 346–347, 355–357
  in reservoirs
    behind dams, 144
    pumped storage, 164–166, 168

Index

Fish populations (*continued*)
　in tailwaters, 139
Fish and Wildlife Service. *See under*
　U.S. Government, Department of
　the Interior
Fission products, 25, 27
　released from reactors, table, 36
Florida, 375
Florida Power and Light Company,
　109, 382
Fluidized-bed combustion, 189–213
　carbon-lean ash-agglomerating, 204
　carbon-lean (BCURA) high-pressure, 193–199
　　figure, 193
　carbon-rich, 198–199, 207–209
　　figure, 199
　carbon-rich ash-agglomerating, 200–203
　　figures, 200, 202
　CERCHAR, 200–203
　　figure, 200
　circulating (Lurgi), 201–203
　in Coalplex, 209–213
　cost reductions from, 195–196
　figure, 190
　with gas turbine for combustion gases, 195–198
　Ignifluid boiler for, 191–193
　　figures, 191–192
　of low-sulfur coke, 211
　multiple-zone, 211–212
　Szikla-Rozinek, 205–206
Fluidized bed for absorption of SO_2
　from coal combustion, 206–208
　from oil combustion, 238
Fly ash, 107–123
　biological availability, 111
　effect of combustion temperature on, 191–192
　electrical resistivity, 183
　quantities produced, 109–110, 112, 181
　radioactivity of, 116
　table, 108
Fly ash removal
　cost of control equipment, figure, 186

　electrostatic precipitator for, 183–186
　　figure, 183
　　photograph, 184
　by scrubbing, 189, 204 (*See also* Scrubbing)
FMC Corporation, 208, 210
Foerster, J. W., 95
Fontana dam, photograph, 134
Formaldehyde in smog, 278–279
Fossil fuel. *See* Fuels
FPC. *See under* U.S. Government
Franklin Institute Research Laboratories, 405–406
Freezing for acid mine water, 334–335
Fuels. *See also* Coal; Gas; Oil
　carbon dioxide from combustion of, 247
　consumed for energy production, 249
　in U.S., 228–229
　nitrates from combustion of, 293–300
　nuclear, 23–24, 75
　future availability of, 214
　Maryland policy regarding, 95
　reprocessing of, 47
　radioactive fly ash content of, table, 108
　selection of, 18
　sulfur content of, 228, 307–309
　　New Jersey regulation on, 175, 206, 228
　　standards on, 228
　　table, 309
Furnace types for coal combustion, 181

Gas
　produced from coal, 209–213
　replacement for coal, 15
Gas turbine
　and coal gasification, 205–206
　for expansion of combustion gases, 195–196
　input gas temperature of, 197–198
　and oil desulfurization at power plant, 238–240

Index

Gas turbine (*continued*)
 for peaking capacity, 158, 162, 196
Gas turbine topping in steam plants, 196–198
 cost reductions from, 195–196
Gaseous reactor waste
 activation and noble gases, tables, 115
 radionuclides released, table, 29, 36
Gasification
 of coke, 179
 Lurgi gasifiers, 205–206
Generators, useful life of, 12–13
Genes, radiation damage to, 54–58, 68–71, 96–99
 tables, 98–99
Genetic death, 97–99
Genetic handicap, order of risk of, 99–101
 compared with other risks, figure, 100
Geological Survey. *See under* U.S. Government, Department of the Interior
Georgia Power Company, 348
Germ cells, female, 82–83
Gestation, effect of radiation on, 83–85
Ghana, 143, 151–152
Godel, Albert, 191–193, 199–200
Gofman, John W., 53
Gonads, female fetal, 82
Great Lakes, 383
Greenhouse effect, 251–252, 284
Ground water near reservoirs, 138, 142, 146
 of pumped storage type, 168–169
Guidelines for radiation protection. *See* Radiation protection standards
Gulf Oil Company, 234–235

H-3. *See* Tritium
H-Oil process, 234
Half-life, of reactor gaseous wastes, table, 29
Halogens, in gaseous reactor waste, table, 36

Harward, Ernest D., 107–125
Heat balance. *See* Energy balance, earth
Heat pollution, atmospheric, 254–257. *See also* Waste heat
 from power plants, table, 398
Heat rate. *See* Thermal efficiency of power plants
Heat rejection methods, 399–405
 flow diagrams, 402
 table, 403
Hematopoetic organs, cancer induced in, 50
Hemoglobin, fish, oxygen affinity in heated water, 356
HEW. *See under* U.S. Government
Hickel, Walter J., 408
Hoover dam, 164
Hudson River, 161
Humble Oil, 235
Humboldt Bay power plant
 noble and activation gases from, table, 115
 radioactive effluent, tables, 36
Hydrocarbon Research, Inc., 234–235
Hydroelectric power generation
 dams, environmental effects, 133–154
 figure, 137
 photographs, 134, 140, 141, 143, 151
 installed capacity, 159
 potential capacity, U.S., 159, 161
 pumped storage, 158–171
  environmental effects, 164–169
  photographs, 160, 165
  size of equipment, 166
Hydrogen. *See also* Tritium
 in smog, 275
 in stratosphere and mesosphere, 280–281
Hydrogen sulfide, 179–180, 192, 207–213, 237–238
 atmospheric reactions of, 269
 conversion to sulfur, 207–209, 238
 in reservoirs, 144
 in stratosphere, 281

ICRP. *See* International Commission on Radiological Protection

IdimetsuKosan, 235
Ignifluid boiler, 191–193, 199–200
  figures, 191–192
Illinois, 374
  reclaimed parkland in, photograph, 332
  strip mining in, 326
Illinois River, 374
Indian Point power plant, figure, 26
  noble and activation gases from, table, 115
  radioactive effluent from, tables, 36
Institut Français du Pétrole, 234
Insurance against nuclear accidents, 127–128
International Atomic Energy Agency (IAEA), 129
International Biological Program (IBP), 153
International Commission on Radiological Protection (ICRP), 30, 32, 40, 65, 78, 98–99, 101, 103, 109, 112, 114, 129
International Council of Scientific Unions (ICSU), 153
Iodine-131, 64, 109
  buildup in milk, 32–33
  in food chain, 64
  in liquid reactor waste, 28
  maximum permissible concentration in air, 33, 77
Ion exchange for acid mine water, 334–335
Isotope ratio method
  sensitivity of, 304–305
  for sulfur, 302–315
Isotope ratio of sulfur in fuel, table, 309
Ivory Coast, 152

Jaske, Robert T., 9, 387–393
Jéquier, L., 200–203
Jersey Central Power and Light Company, 160, 163, 313
Johns Hopkins University, The, 56–57, 69–70, 374
Joint Committee on Atomic Energy (JCAE). See under U.S. Government, Congress

Kansas Power and Light Company, 189, 241
Kashima, 235
Kellermann Power Station, 205
Kentucky, strip mining, 319, 326
  photographs, 319, 321
Kentucky dam, photograph, 141
Keystone Generating Station, 311
  photograph, 313
Kossou dam, 152
Krypton-85, 37, 109
  atmospheric accumulation of, 37, 81
  projected content, 121
  half-life, 29
  in reactor waste, 29
Krypton-87, in reactor waste, 29
Krypton-88, in reactor waste, 29
Kuwait crude oil, sulfur content, table, 309
Kuwait National Petroleum, 235

La Crosse power plant, radioactive effluent, tables, 36
Lagler, Karl F., 133–157
Lake Mead, 164
Lake Nasser, 138
Lakeside Station steam plant, 180
Land reclamation, 329–332
  aerial seeding for, photograph, 330
  of mine drainage areas, 332–336
    cost, 334–336
  of refuse banks, 325–326
  of strip mines, 327–329
    cost, 327–329
  renewed area, photographs, 331–332
  for wildlife habitat, 337–338
Landscape changes, effect on climate, 254–255, 260–261
Landscape pollution. See Pollution, landscape
Laster, Howard, 95
Laughnan, John R., 63–74
Lawrence Radiation Laboratory, 45
Lawrence Station power plant, 189, 241

Index

Leukemia, radiation-induced, 50–52, 65–67, 130
  in mice, 67
  by neutrons, 68
Libya, low-sulfur oil exploration, 228–229
Licensing
  of hydroelectric projects, 166
  of nuclear reactors, 30–33, 77, 127–128, 372
Light water reactors. *See* Nuclear reactors, boiling water, pressurized water
Lignite, global consumption, 249
Limestone, for sulfur oxide removal
  in fluidized bed, 206–208
  in PF furnace, 188, 241
Liquid natural gas (LNG), to replace coal, 15
Liquid reactor waste
  in hypothetical river, 45–48
  discussion of river, 75–76
  radionuclides in, table, 28
Long Island Lighting Company (LILCO), 312, 314–315
Long Island Sound, 163
Los Angeles 274–276
Louisville Courier-Journal, 319, 321
Luce, Charles F., 365, 408
Lung radiation exposure
  from air near power plant, table, 111
  from atmospheric radon, 31–32
Lung cancer
  radiation-induced, 50
  smoking or radiation, 79
Lurgi circulating fluidized bed, 201–203
  figure, 202
Lurgi gasifiers, 205–206
Lymphatic organs, radiation-induced cancer in, 50–51

McAllister, J., 143, 151
MacDonald, Gordon J. F., 246–262
MacLean, David, 128, 259
Magnetohydrodynamic (MHD) power generation, 204, 217–218
  nitrogen oxides from, 217

Maine, 382
Malaria, 145
Male-to-female birth ratio, radiation effect on, 56–57, 69–71, 85–91
  table, 87
Manganese-54, 75
  maximum permissible concentration, 35
Manhattan District, 34
Man-made lakes, 135
Martin, James E., 107–125
Maryland, 94–95
Maryland Academy of Sciences, 94–106
Massachusetts, 374
Mauna Loa Observatory, 247
Maximum permissible concentration (MPC), 35
  argument for consideration of, 80
  buildup considered in setting, 63–64
  buildup not considered in setting, 44
  exceeded, 47
  as secondary radiation standard, 44–48
Maximum permissible concentrations
  antimony-125, 75
  cesium-137, 46
  iodine-131, 33, 77
  manganese-54, 35
  radium-226, 35
  ruthenium-106, 46
  strontium-90, 46–47
  tritium (H-3), 35
Maximum permissible dose, 31–32
  in United Kingdom, 103
Medical use of radiation, 49, 51–52, 96, 103
  during pregnancy, 50, 69, 84–91
  table, 87
  lack of standards in U.S. for, 96
  order of risk from, 101
Mekong River, 153
Meiosis, 83
Meramec Station power plant, 189, 241
Merz, Timothy, 69–71, 82–93
Mesosphere, 264, 279–284

Metabolism and thermal pollution, 346, 355–356
Methane, atmospheric, 273
 in smog, 275
 in stratosphere and mesosphere, 283–284
Metropolitan Edison Company, 187
Meyer, Mary B., 69–71, 82–93
MHD. *See* Magnetohydrodynamic power generation
Miami, University of, 376
 Institute of Marine Science, 382
Michigan, 348, 381
Middle East oil, 231, 234–235
 sulfur content of, table, 309
 typical properties of, table, 231
Migration of fish
 past dams, 146–147
 past pumped storage plants, 168
 and thermal pollution, 355–357, 373
Mines. *See* Coal mining
Mississippi River, 374, 389
Mizushima, 235
Monsanto Chemical Company, 187
Montana crude oil, sulfur content of, table, 309
Morgan, Karl Z., 103n
Morgantown power plant, 375
Muddy Run pumped storage plant, 164, 167–168
 photograph, 165
 pumping rate of, 167
Mutation frequency, 83–91

National Academy of Engineering, 241–242
National Academy of Sciences, 335
National Air Pollution control Administration (NAPCA). *See* under U.S. Government, Department of Health, Education, and Welfare
National Center for Atmospheric Research (NCAR), 266–267
National Coal Association, 332
National Council on Radiation Protection and Measurements (NCRP), 30, 32

National Technical Advisory Committee on Water Quality Criteria, 347, 359
National Technical Advisory Subcommittee for Fish, Other Aquatic Life, and Wildlife, 359
Natural background radiation
 biological effects of, 130
 dose levels from, 31, 95–96
Natural draft cooling tower, 400–403
Natural gas
 converted from coal, 209–213
 for replacing coal, 15
Neafra crude oil, sulfur content, table, 309
Neuroblastoma, radiation-induced, 51
Neutralization of acid mine waters, 333–334
New Haven, SO_2 pollution in, 311
 photograph, 312
New Jersey standards on sulfur in fuel, 175, 206, 228
New York
 Finger Lakes, 354
 Rule 200 on sulfur in fuel, 228
Nickel, 231–232
 in oils, table, 231
Nile River valley, 138, 143
Nitrogen compounds
 atmospheric, 271–273, 297
 fixation by fuel combustion, 297
 in precipitation, 298
 in stratosphere and mesosphere, 281
Nitrogen oxides
 arranged by increasing valence
  nitrous oxide (N_2O), 271–272, 281
  nitric oxide (NO), 272, 276–278, 281
  nitrogen dioxide (NO_2), 272–273, 276–279, 281
  nitrogen pentoxide (N_2O_5), 273, 281
 atmospheric, 294–295, 297–300
 in MHD power generation, 217
 in photochemical smog, 275
 daily variation of, figure, 276
Noble gases released from reactors, 113
 tables, 36, 115
 whole body exposure from, 114

Index   427

Nondegradation, water quality standard, 347
Nondisjunction, 71, 89–90
Norris dam, photograph, 140
North African low-sulfur oil, 232
  exploration for, 228–229
Northeast Utilities Service Company, 161
Northfield Mountain Pumped Storage Project, 161–163, 166–167
  pumping rate for, 167
Northport power plant, 311–315
  photograph, 314
  stack gas plume measurements
    figure, 314
    table, 315
Norway, precipitation, 292–299
  nitrates in, figure, 298
  sulfur in, figures, 293, 296
  total excess acid and base in, figure, 300
Nuclear fuel, future availability, 213–214
Nuclear fuel-reprocessing plant, 47
  Krypton-85 released from, 121
  Maryland policy regarding, 95
Nuclear power plants. *See also* Nuclear power plants, radioactive effluents; Power plants, nuclear
  accident probability, 127–128
  cooling water needs, 344–346, 352–353, 368–369
  efficiency of, 213–214, 344n, 353, 368–369, 398
  fast breeder reactor for
    projected use, 15–16
    uncertainties of, 214
  generating capacity, growth of, 366–367
    figure, 395
  siting of, 127–128
Nuclear power plants, radioactive effluents, 25–29, 34–38, 40, 75–77, 112–123
  accidental release of, 24
  BWR and PWR compared, 116–120
    tables, 118, 120
  compared with fossil plants, 107–123

  from Dresden 1, dose rates, table, 114
  gaseous
    activation and noble gases in, tables of, 36, 115
    radionuclides in, table, 29
    limits on radioactivity of, 30–33 (*see also* Standards for radiation protection)
  liquid
    in hypothetical river, 45–48, 75–76
    radionuclides in, table, 28
    purification of, 27
    storage of
      permanent, 27
      temporary, 29, 116, 120
Nuclear reactors
  boiling water (BWR), figure, 27
  cesium-137, inventory of, 47, 77
  fast breeder, 15–16, 214
  fuel composition and assembly, 23–24
  licensing, 30–33, 77, 127–128, 372
  number in use, 23
  present-day devices inefficient, 213–214
  pressurized water (PWR), 25
    exploded view, figure, 26
  regulations for exporting, 128–130
  temperature coefficient of reactivity of, 24
Nuclear weapons, radioactivity released from, 35, 76

Oak Ridge National Laboratory (ORNL), 23
Oakley, Donald T., 107–125
Oceans
  carbon content of, 247–248, 251
    figure, 248
  oil pollution of, 254–255
  sulfur content of, 290
  and use of waste heat for upwelling, 407
Ohio, strip mining in, 326
  reclaimed land, photograph, 331
Ohio River, 389

Index

Oil
  consumed for energy production, 249–250
  conversion from coal, 209–213, 215
  depletion allowance, 231
  low-sulfur
    exploration for, 228
    use in replacing coal, 15
  market for high sulfur content, 228
  radioactive fly ash content of, table, 108
  reserves of, 250
  sulfur content of, table, 309
    and cost factors, 233–237
    standards for, 175, 206
Oil desulfurization
  in combustion
    complete, 237–238
    partial, 238–239
  cost of, figure, 236
  by cracking, 239–240
    figure, 239
  at power plant, 235–242
  at refinery, 231–235
Oil pollution, oceans, 254–255
Oklahoma crude oil, sulfur content of, table, 309
Olefins, in smog, 275, 277–279
Olympic peninsula, 382
Once-through cooling, 346, 348–349, 353, 369–370
  alternatives to, 400–401
  reducing effects of the discharge, 399–400
Onchocerciasis, 146
Oocytes, radiation damage to, 71, 83
Oogonia, radiation damage to, 83
Oregon, 349
  Eugene Water and Electric Board, 382
Oroville dam, 163
Oxygen
  atmospheric, man-induced changes in, 261
  in cooling waters and streams, 346, 355–356
  reactions of
    in stratosphere and mesosphere, 279–284
    in troposphere, 266, 269–279
    in reservoirs and tailwater, 139, 144, 166
Ozone, 269, 272
  in photochemical smog, 269, 275–278
  daily changes in, figure, 276
  radiation absorption of, 248, 266
  in stratosphere and mesosphere, 279–284

PAN, in smog, 278–279
Pancreas cancer, radiation-induced, 50
Paraffins, atmospheric, 273
  in smog, 275
Particulates released from reactors, table, 36
Paulson, Glenn L., 18, 130
Peach Bottom Atomic Power Station, 164
  radioactive effluent, 36
Pelvimetry, during pregnancy, 50, 69, 84–91
  and reproductive performance, 82–91
  table, 87
Pennsylvania
  cleanup of acid mine drainage, 335–336
  refuse banks in, 323–326
    photographs, 324, 326–327
  strip mining, 326
  subsidence in, 322–323
    photograph, 322
Pennsylvania Power and Light Company, 313
Periphyton, 144
Peroxy compounds and radicals in smog, 277–279
Perry, Harry, 317–339
Petroleos Mexicanos, 235
Petroleum, global reserves, 250. See also Fuels; Oil
PF. See Pulverized-fuel combustion
Philadelphia Electric Company, 165, 313

Index

Phillips, Owen, 95
Photochemical reactions, 264–266, 268, 272, 290
  in polluted air, 274, 276
  in stratosphere and mesosphere, 279–284
Photochemical smog, 269, 272, 274
  typical composition of, table, 275
  daily variations in, figure, 276
Photochemical yield, 270
Photolysis, 279–281
Photosynthesis, role of carbon dioxide in, 247, 251
  figure, 248
Phytoplankton, 144
  and agricultural nutrient in oceans, 251
Pintsch-Bamag, 208
Plankton, 144
Pole Station, 247
Pollution. *See also* Pollution, atmospheric; Pollution, landscape; Pollution, radioactive; Pollution, thermal; Pollution, water
  approaches for abatement, 4
  control through regulation and subsidy, 9–10
  perception of, by man, 8
Pollution, atmospheric, 274–279
  composition of, table, 275
  daily variations in, figure, 276
  nitrogen oxides in, 294–295, 297–300
  organic compounds in, 273
  polluted precipitation, 292–299
    total excess acid and base in, figure, 300
  sulfur in, 302
  sulfur dioxide in, 289–299
Pollution, landscape, 317–338
  acid mine drainage, 320, 329–336
  agricultural, 254–255, 260–261
  deforestation, 260–261
  erosion related to dams, 138, 146, 152
  mine fires, 323
  reclamation of, 329–332 (*see also* Reclamation)
  refuse banks, 323–326
    photographs, 324, 326–328
  strip mining, 319–322, 326–329
    photographs, 319–321
  subsidence, 322–323
    photograph, 322
Pollution, radioactive, 25–29, 34–38, 40, 75–77, 112–123
  accidental, 24
  from Dresden 1, dose rates, table of, 114
  effects of (*see* Biological effects of radiation)
  effects of storage time on, 29, 116, 120
  gaseous reactor waste
    activation and noble gases in, tables, 36, 115
    radionuclides in, table, 29
    limits on, 31–33 (*see also* Standards for radiation protection)
  liquid reactor waste
    in hypothetical river, 45–48, 75–76
    radionuclides in, table, 28
  nuclear and fossil plant releases compared, 107–123
  pathways to man, 45, 113
  permanent storage of, 27
Pollution, thermal
  atmospheric, 398
  control of, 348–350
    table of costs, 349
  cooling water requirements, 344–346, 352, 368, 394–404
    tables, 345, 398, 403
  effects of, 346–347, 357–358
    on aquatic ecology, 353–358
    on fish, 355–357, 373–374
  evaporation losses, 138, 145, 399
  modeling and prediction, 374–375, 389
  from nuclear and fossil plants, compared, 344n, 352–353, 368–369, 397–398
    table, 398
  and power plant size, 345, 353, 394, 396–399
  and public power policy, 388–389
  reduction of, 217

Index

Pollution (continued)
  temperature rise from, 346, 349, 353–354
  uses for, 348, 358, 381–382, 388, 405–408
  water quality standards for, 347–348, 359–360, 372
Pollution, water
  by acid mine drainage, 320, 329–332
  by chemicals in cooling water, 404–405
Pope, Evans, and Robbins, 193
Portland General Electric Company, 349
Potassium-40, 29
Potomac Electric Power Company, 349
Potomac estuary, 375
Pour point for oils, table, 231
Power companies
  Atlantic City Electric, 313
  Baltimore Gas and Electric, 94, 313
  Boston Edison, 241
  Carolina Power and Light, 348
  Commonwealth Edison, 11–12, 14–15, 110, 241
  Consolidated Edison of New York, 161, 182–186, 228–229, 312, 408
  Consumers Power, 348
  Delmarva Power and Light, 313
  Detroit Edison, 188
  Florida Power and Light, 109, 382
  Georgia Power, 348
  Jersey Central Power and Light, 160, 163, 313
  Kansas Power and Light, 189, 241
  Long Island Lighting, 312, 314–315
  Metropolitan Edison, 187
  Northeast Utilities Service, 161
  Pennsylvania Power and Light, 313
  Philadelphia Electric, 165, 313
  Portland General Electric, 349
  Potomac Electric Power, 349
  Public Service Electric and Gas, 160, 163, 313
  UGI, 192, 326
  Union Electric, 169, 189, 241
  United Illuminating, 312
  West Texas Utilities, 196
Power plants. *See also* Power plants, fossil fuel; Power plants, hydroelectric dams; Power plants, nuclear; Power plants, pumped storage; Power plants, siting
  coal-fired, 10–18, 177–182, 213–217
  power and chemicals from, 17, 209
  radioactivity from, compared with nuclear, 107–123
    tables of, 111, 118, 120
  stack height and radiation exposure, 119–120
  cooling water requirements, 344–346, 352, 368, 398
  efficiency of (*see* Efficiency of power generation)
  gas turbines, 195–198
  generators for
    size of, 345, 367, 396–397
    useful life of, 12–13
  oil-fired, 228–242
Power plants, fossil fuel
  Arthur Kill, 312
  English Station, photograph, 312
  Kellermann, 205
  Keystone Generating Station, 311
    photograph, 313
  Lakeside Station, 180
  Lawrence Station, 189, 241
  Meramec Station, 189, 241
  Morgantown, 375
  Northport, 311–315
    photograph, 314
    stack gas plume measurements
      figure, 314
      table, 315
  Ravenswood Station, 182–186
    figures, 182–183
    photograph, 184
  San Angelo Station, 196
  Turkey Point, 109, 111, 116, 375, 382
  Widows Creek, 110–111, 113, 116
    bone and lung dose rates from, 114
    fly ash and particulates from, 116
    radioactive releases and dose rates, table, 111
    site characteristics, 116

Index

Power plants, hydroelectric dams. See also Dams
  Aswan, 138, 143
  Conowingo, 164, 167
  Fontana, photograph, 134
  Hoover, 164
  Kentucky, photograph, 141
  Kossou, 152
  Norris, photograph, 140
  Oroville, 163
  Rampart, 153
  Thermalito, 163
  Turners Falls, 167
  Wheeler, 380
Power plants, nuclear. See also Nuclear power plants
  Big Rock, radioactive effluent, tables, 36, 115
  Browns Ferry, 380
  Calvert Cliffs, 94, 375
  Connecticut Yankee, radiocative effluent, tables, 36, 115
  Dresden, 11, 110, 113–115
    radioactive effluent, tables, 36, 114, 115
    site characteristics, 110
  Elk River, radioactive effluent, tables, 36
  Humboldt Bay, radioactive effluent, tables, 36, 115
  Indian Point, figure, 26
    radioactive effluent, tables, 36, 115
  La Crosse, radioactive effluent, tables, 36
  San Onofre, radioactive effluent, tables, 36, 115
  Saxton, radioactive effluent, tables, 36
  Shippingport, 215
  Shoreham, 120
  Trojan, 349
  Turkey Point, 109, 111, 116, 375, 382
  Yankee, 109
    radioactive effluent, tables, 36, 115
Power plants, pumped storage, 158–171
  Cornwall, 161

Muddy Run, 164, 167–168
  photograph, 165
Northfield Mountain, 161–163, 166–167
San Luis, 163
Taum Sauk, 169
Thermalito, 163
Tocks Island, 160, 163
Yards Creek, 163
  photograph, 160
Power plants, siting, 18–19, 367, 371, 387, 397
  and cooling method, 401–405
  of dams and pumped storage, 161–162, 166
  of nuclear plants, 127–128
  required for future, 395, 397
Power production
  atmospheric heating from, 254–257
  cooling water requirement, 344–346, 352
    table, 345
  dependence of society on, 7–10
  economic factors, 7–10
    cost of thermal pollution control, 349–350
    table, 349
    investment cost for cooling, 370–371, 401–404
    tables, 370, 403
    rates, 7–10
  global, 249
  per capita, 256–257, 366
  large vs. small plants, 12
  and national policy, 391–392
  U.S., 3, 12, 16, 249, 351, 366–367, 394–395
    in Boston-Washington megalopolis, 390
    figures, 390, 395
    installed capacity, 394–395
    skyscraper requirement, 391
    tables, 344, 366
Precipitation, 292–299
  nitrates in, figures, 298
  sulfur in, figures, 293, 296
  total excess acid and base in, figure, 300

Precipitators, electrostatic, 13, 183–186
  cost of, figure, 186
  design curves for, figure, 185
  operating temperature, 183
  performance for low-sulfur coal, 183
  size of
    figure, 183
    photograph, 184
Pressurized water reactor (PWR). *See* Nuclear reactors, pressurized water
Price-Anderson Act, 127–128
Primary standard for radiation dosage, 44, 48. *See also* Standards for radiation protection
Producer gas, historical use and production, 179, 205
Public Service Electric and Gas Company, 160, 163, 313
Pulverized-fuel (PF) combustion
  boilers for, 181
    schematic views, 182–183
  history of, 177–182
  time for a change from, 183–191
Pumped storage, 158–171
  advantages of, 161–163
  cost of, 162
  efficiency of, 159
  environmental effects of, 164, 166–169
  installed capacity of, 161
  for peak-load requirements, 158
  photographs, 160, 165
  pumping rates, 167–168
  and water management, 162–164
Purification of reactor coolant, 27

Quabbin reservoir, 163

Radiation
  biological effects of
    on birth rate, 85
    disease, 49–52, 64–68
    on fetus, 50–51, 53–58, 62–72, 85–87
    genetic, 54–58, 68–71, 96–99
    on oocytes and oogonia, 71, 83
    on reproductive performance, 83–84
    table, 87
    superovulation, 83, 88
  doubling dose, 50–52
    discussion on use of, 64–66, 80
    proposed values for, table, 53
    table, 50
  natural background
    biological effects of, 130
    dose levels from, 31, 95–96
Radiation balance for earth, 255–257, 260–261
  effect of gaseous pollutants on, 248–249
  figure, 256
  need for further study of, 257–259
Radiation damage
  repair of, 68
  risk of, compared to other risks, 97–101
Radiation dose
  to bone from radionuclides in water 46–47, 75–76
    table, 46
  to bone and lung from air near power plant, table, 111
  lethal, for oocytes, 84
  from medical exposure, 49, 96
  from natural sources, 31, 95–96
  threshold, 53–54, 68, 79
  voluntary and involuntary, 96–97, 101
Radiation protection, safety record, 33–34
Radiation protection standards, 31, 44, 95, 97, 102–103
  criticized, 44–48
  favored, 63–65, 78–79
  federal or state, 126
  genetic effects not overlooked in setting, 71–72
  genetic effects overlooked in setting, 55
  maximum permissible concentrations (MPC), 35, 44–48, 80
  buildup factors considered in setting, 63–64

Index 433

Radiation protection standards, maximum permissible concentrations (*continued*)
  buildup factors not considered in setting, 44
  for occupational exposure, 34, 78
  proposed reductions of, 58, 102–103
  effect of power plants, 126
Radioactivity released to environment, 25–29, 34–38, 40, 75–77, 112–123
  accidental, 24
  from Dresden 1, dose rates, table of, 114
  effect of storage time on, 29, 116, 120
  as gaseous reactor waste
    activation and noble gases, tables of, 36, 115
    radionuclides in, table, 29
  limits on, 31–33 (*see also* Standards for radiation protection)
  as liquid reactor waste
    into hypothetical river, 45–48, 75–76
    radionuclides in, table, 28
  nuclear and fossil releases compared, 107–123
  pathways to man, 45, 113
  permanent storage of, 27
Radium-226, 29, 35n
  in air near power plant, table, 111
  in fly ash, 108–109
  maximum permissible concentration, 35
  power plant release rate, 112
Radium-228, 29
  in fly ash, 108
  power plant release rate, 112
Radon, 31–32
Rampart dam, 153
Rate structure, economic factors, 7–10
  cost of thermal pollution control, 349–350
  table, 349
  investment cost for cooling, 370–371, 401–404
  tables, 370, 403

Ravenswood Station power plant, 182–186
  figures, 182, 183
  photograph, 184
Reactor. *See* Nuclear reactors
Reclamation, 329–332
  aerial seeding for, photograph, 330
  of mine drainage areas, 332–336
    cost, 334–336
  of refuse banks, 325–326
  and rehabilitation, 329
    photographs, 331–332
  of strip mines, 327–329
    cost, 327–329
  renewed areas, photographs, 331–332
  for wildlife habitat, 337–338
Refuse banks, 323–326
  on fire, 325
    photographs, 327–328
  photographs, 324, 326
  reclamation of, 325–326
Regulatory agencies for nuclear safety, 30–33
Rejected heat. *See* Waste heat
Reproductive performance, effect of radiation on, 83–91
Reservoirs, 135, 138–139, 142–145, 150
  pumped storage, 163–169
Residence time, atmospheric
  of sulfur, 289–290
  of nitrogen oxides, 297
Residual fuel oil, 228–242. *See also* Oil
Reverse osmosis for acid mine waters, 334
Rheumatoid spondylitis, 49
Rich, Robert P., 95
River Bend crude oil, sulfur content, table, 309
River blindness, 146
Rivers
  damming of, 136, 138
  flooding stage, 146, 150
  flow rates and pumped storage, 167–168
  flow stabilization, 138, 146

Index

Rivers (*continued*)
  and liquid reactor waste, 45–48, 75–76
  table of radionuclides, 28
Roberts, Mary C., 260
Ruthenium-106, 75
  maximum permissible concentration, 46

Safoniya crude oil, sulfur content of, table, 309
San Angelo Station power plant, 196
San Joaquin valley, 163
San Luis dam, 163
San Onofre power plant
  noble and activation gases from, table, 115
  radioactive effluent, tables of, 36
Sand oil, 250
Saudi Arabia crude oil, sulfur content, table, 309
Saxton nuclear power plant, radioactive effluent, tables, 36
Scandinavia, precipitation, 292–299
  nitrates in, figure, 298
  sulfur in, figures, 293, 296
  total excess acid and base in, figure, 300
Schistosomiasis, 145
Scrubbing, 14, 187, 240–241
  with calcined stone in water, 188–189, 241
  of fuel gas for hydrogen sulfide removal, 207–209
  need for alternatives to, 187, 240–242
  for fly ash removal, 189, 204
  for sulfur oxide removal
    from fluidized bed gases, 207
    from stack gases, 188–189, 240–241
Seaborg, Glenn, 4
Secondary radiation standards, 44
Sedimentation in reservoirs, 138
Seismic effects of dams, 141
Seliger, Howard H., 95
Settling basin, reservoir acting as, 138
Sex ratio at birth, radiation effects on, 56–57, 69–71, 82–91
  table, 87

Shale oil, 250
Shippingport nuclear power plant, 215
Shoreham nuclear power plant, 120
Sho-Vel-Tum crude oil, sulfur content of, table, 309
Signal Oil Company, 235
Singer, S. Fred, 341–350
Site selection for power plants, 18–19, 367, 371, 387, 397
  and cooling method, 401–405
  dams and pumped storage, 161–162, 166
  nuclear, 127–128
  number required, 367, 395
Smith, Frederick E., 7–10
Smog, 268–269, 274–275
  photochemical, 269, 272
  daily variation in, figure, 276
  typical composition of, table, 275
Socolar, Sidney J., 130
Soil stabilization, 328
Solar constant, 256
Solar radiation, 266
  and photochemistry, 272, 274, 276
  in stratosphere and mesosphere, 279–284
  spectral intensity distribution, figure, 268
Somatic effects of radiation, 65–66, 68, 101
Spawning in heated waters, 356–357
Specific gravity of oils, table, 231
Spondylitis, 49, 65
Spray ponds, 348, 400–404
  flow diagram, 402
  investment costs, table, 403
Squires, Arthur M. (coeditor), 175–245
Stack gas cleanup, 14, 187, 240–241. *See also* Scrubbing
Standards for radiation protection, 31, 44, 95, 102–103
  criticized, 44–48
  favored, 63–65, 78–79
  federal or state, 126
  genetic effects not overlooked in setting, 71–72
  genetic effects overlooked, 55

Standards for radiation protection (*continued*)
  maximum permissible concentrations (MPC), 35, 44–48, 80
    buildup factors considered in setting, 63–64
    buildup factors not considered, 44
    for occupational exposure, 34, 78
    proposed reductions of, 58, 102–103
    effect on power plants, 126
Standards for thermal pollution of water, 347–348, 359–360, 372
Stapleton, George E., 63–74
Starr, Chauncey, 126
Steigelmann, William H., 394–411
Steinberg, Meyer, 302–316
Steinkohlen-Electrizität AG, 205
Sterility in irradiated animals, 83
Stewart, O. W., 18
Stomach cancer, radiation-induced, 50, 65
Storage of reactor waste
  permanent, 27
  temporary, 29, 116, 120
Storm King, 161
Stratopause, 264, 266
Stratosphere, 264–265, 269, 271–272, 279–284, 289
Strip mining, 319–322, 326–329
  and acid mine drainage, 329–330
  coal production from, 317–318
  sulfur content of, 318
  cost of reclamation, 327–329
  photographs, 319, 320, 321
Strontium-89, in liquid reactor waste, table, 28
Strontium-90
  from reactors, 38
  in liquid waste, 75–77
    table, 28
  maximum permissible concentration, 46–47
  from weapons tests, 35, 38
Study Panel on Nuclear Plants, Maryland Academy of Sciences, 94–106
Subsidence in coal mine regions, 322–323

Subsidies for pollution control, 9
Suess effect, 120
Sulfur
  atmospheric, 268–271, 289–300, 302
  reactions of, 269–271
  sources of, 268, 289
  as by-product of stack gas, 187
  in fuel, 228
    coal, 307
    cost factors of, 233–237
    New Jersey regulations on content of, 175, 206, 228
    oil, 307–308
    table, 231, 309
  as hydrogen sulfide from incomplete combustion, 179–180, 192, 207–213, 237–238
  isotope composition, table, 303
  isotope ratio, in fuels, 309
  del values for, table, 303
  natural circulation of, 290
    figure, 291
  in precipitation, 290–294
    figures, 293, 295–296, 300
  recovered from calcium sulfide, 207–209, 238
  recovered in Coalplex, 209–213
  released by man, 290
  removal at power plant, 187–189, 235–242
    cost alternatives, 236
  removal at refinery, 231–234
    catalytic hydrodesulfurization, figure, 232
    plants for desulfurization, table of, 235
  in stratosphere and mesosphere, 281, 283
Sulfur dioxide (SO_2)
  atmospheric, 289–299
  in combustion gases, 186
  conversion to sulfer trioxide in atmosphere, 271, 283, 308–315
  cost of control, 175–176
  environmental effects of, 41, 302
  reaction with half-calcined dolomite, 218

Sulfur dioxide (*continued*)
removal (*see also* Scrubbing)
in fluidized bed, 206–208, 218–219, 238
in oil combustion, 238
from stack gas, 14, 187–189, 240–241
in smog, 275
in stratosphere, 281, 283
yearly discharge of, 175
Sulfur hexafluoride, as tracer, 312, 315
Superovulation, radiation-induced, 83, 88
Susquehanna River, 164, 167
Sweden, precipitation, 292–299
nitrates in, figure, 298
sulfur in, figures, 293, 296
total excess acid and base in, figure, 300
Swimmer's itch, 145

Tamplin, Arthur, 44–60, 63–72, 75–77, 79–80, 82, 90
Taum Sauk pumped storage project, 169
Temperature
atmospheric, 264
figure, 265
historical changes in, 253
in estuaries and coastal zones, prediction of, 374–375
Temperature of cooling water, 346, 349, 353–354
effects on aquatic ecosystems, 354–357
fish, 355–357, 373–374
other effects of, 357–358
Tennessee, contour strip mining, photograph, 320
Tennessee River, 140–141
Thermal balance, 255–257, 260–261
and air-water energy transfer, 254–255
effect of carbon dioxide, 251–254
effect of gaseous pollutants, 248–249
figure, 256
need for study of, 257–259
perturbed by man, 255–257

Thermal effects in water, U.S. Government research programs on, 376–383
biological effects, 378–380
nontreatment solutions, 381–382
plant site selection, 383
transport and behavior of heat in water, 377–378
treatment processes, 380–381
Thermal efficiency of power plants
with bottoming cycles, 406–407
figure, 391
with gas turbine or MHD topping, 217
nuclear, 213–214, 398
nuclear and fossil compared, 344n, 353, 368–369, 397–398
pumped storage, 159
relation to cooling cycle, 396
and siting requirements, 387–388
Thermal energy released by man, 249
effect on heat balance, 254–257
Thermal pollution
atmospheric, 398
control of, 348–350
table of costs, 349
cooling water requirements, 344–346, 352, 368, 394–404
tables, 345, 398, 403
effects of, 346–347, 357–358
on aquatic ecology, 353–358
on fish, 355–357, 373–374
evaporation losses, 138, 145, 399
modeling and prediction, 374–375, 389
from nuclear and fossil power plants, compared, 344n, 352–353, 368–369, 397–398
table, 398
and power plant size, 345, 353, 394, 396–399
and public power policy, 388–389
reduction of, 217
temperature rise from, 346, 349, 353–354
uses for, 348, 358, 381–382, 388, 405–408

Index

Thermal pollution *(continued)*
 water quality standards for, 347–348, 359–360, 372
Thermal stratification in reservoirs, 139
Thermalito dam, 163
Thermosphere, 264
Thomas, William A., 126
Thorium, 29, 107
Thorium-228, in fly ash, 108, 112
Thorium-230, 112
Thorium-232
 in air near power plant, table, 111
 in fly ash, 108–109
 power plant release rate, 112
 and reactor efficiency, 213–214
Threshold dose, 53, 68
 concept challenged, 54
 explained, 79
Thymic enlargement, 49
Thyroid cancer, radiation-induced, 50, 66
 order of risk, 97
Tittabawassee River, 348
Tocks Island, 160, 163
Tracer methods for sulfur, 302–315
Transmission lines, extra-high-voltage (EHV), and power plant siting, 18–19
Tritium
 amount released from reactors, 28–29, 35–38, 81
 table, 36
 and genetic mutation, 38
 half-life of, 29
 in hydrosphere, 37–38
 maximum permissible concentration, 35
Trojan nuclear power plant, 349
Tropopause, 264, 266
Troposphere, 264–279
 polluted, 274–279
Tumors, 66
 radiation-induced, tables, 50–51
Turkey Point power plants, 109, 111, 116, 375, 382
Turners Falls reservoir, 167
TVA. *See under* U.S. Government

UGI Corporation, 192, 326
Union Electric Company, 169, 189, 241
Union Oil Company, 234
United Aircraft, 206
United Illuminating Company, 312
United Kingdom Atomic Energy Authority, 103
United Nations, 129
 U.N. Development Program (UNDP), 153
 Food and Agriculture Organization (FAO), 153
 Scientific Committee on the Effects of Atomic Radiation (UNSCEAR), 30, 77
United States Government
 Advisory Committee on Reactor Safeguards (ACRS), 31
 Atomic Energy Commission, 4, 5, 48, 102, 378, 382–384, 409
  dual role as developer and protector, 39–43, 62, 81
  and export of nuclear technology, 128–130
  licensing and regulation, 30–33, 77, 127–128, 372
  Pacific Northwest Laboratory, 374
  safety record, 33–34, 38–39
  thermal pollution, position on regulation of, 372
  waste heat research program, 372–376, 389
 Congress
  Joint Committee on Atomic Energy (JCAE), 11, 42, 127
  Clean Water Restoration Act, 387
  Federal Water Pollution Control Act, 347
  Price-Anderson Act, 127–128
 Department of Agriculture, 329
 Department of Defense, 5
 Department of Health, Education, and Welfare (HEW), 13, 15, 30
  Bureau of Radiological Health, 31, 107, 110

Index

United States Government, Department of Health, Education, and Welfare (*continued*)
  National Air Pollution Control Administration (NAPCA), 206, 219
  Public Health Service, 31
  role in reactor regulation, 39
Department of the Interior, 335, 347, 351, 372–373
  Bureau of Commercial Fisheries, 372, 382
  Bureau of Land Management, 343
  Bureau of Mines, 193, 208, 217–218, 323, 329, 343
  Bureau of Outdoor Recreation, 343
  Bureau of Reclamation, 343
  conservation and development activities of, 343
  Federal Water Quality Administration (FWQA), 329, 372, 378, 380–381, 409
  Fish and Wildlife Service, 343, 351
  Geological Survey, 343
  National Park Service, 343
  Office of Coal Research, 193, 209, 217–218, 343, 406
  Office of Saline Water, 343
  Regional Power Administration, 343
  Secretary of the Interior, 408
    as protector of nation's water, 347
Department of Justice, 372
Federal Power Commission (FPC), 159, 161, 170, 349, 383, 409
  licensing of hydroelectric projects, 166
Federal Radiation Council (FRC), 30, 40, 48–49, 95–96, 102
National Academy of Engineering, 241–242
National Academy of Sciences, 335
Oak Ridge National Laboratory, 23
Office of Science and Technology, Committee on Water Resources Research, 377

Tennessee Valley Authority (TVA), 110, 188, 320, 380
  photographs, 134, 140, 141
  trials of limestone injection for SO_2 removal, 188, 241
Water Resources Council, 368
United States Government policies
  power policy, 387–388, 391–392
  program in global energy balance, 257–259
  response to conflicting demands, 343
  role in pollution control, 9–10
United States Government research program on thermal effects in water, 370–383
  biological effects, 378–380
  nontreatment solutions, 381–382
  plant site selection, 383
  transport and behavior of heat in water, 377–378
  treatment processes, 380–381
Universal Oil Products, 234–235
Upwelling, use of waste heat for, 407
Uranium, 24, 29, 107
Uranium-235, and reactor efficiency, 213–214
Uranium-238
  in air near power plant, table, 111
  and reactor efficiency, 213–214
  release rate, 112
Uranium oxide, reactor fuel, 23–24, 75
Utility companies. *See* Power Companies

Vanadium, 231–232
  in oils, table, 231
Venezuelan oil
  metals content of, 231–232
  typical properties, table, 231
  use in U.S., 229–231, 233–234
Vermont, 352
Volcanic dust, effect on climate, 260
Volta Lake, 152
  photographs, 143, 151

Warren, Shields, 78–81
Washington (state), 373, 382

Index

Waste, from nuclear reactors
  accidental release of, 24
  BWR and PWR compared, 116–120
    tables, 118, 120
  compared with fossil plants, 107–123
  from Dresden 1, dose rates, table, 114
  gaseous
    activation and noble gases, tables, 36, 115
    radionuclides released, table, 29
    limits on radioactivity of, 30–33 (*see also* Standards for radiation protection)
  liquid
    in hypothetical river, 45–48, 75–76
    radionuclides released in, table, 28
  purification of, 27
  storage of
    permanent, 27
    temporary, 29, 116, 120
Waste heat
  to atmosphere, 398
  control of, 348–350
    table of costs, 349
  and cooling water requirements, 344–346, 352, 368, 394–404
    tables, 345, 398, 403
  effects of, 346–347, 357–358
    on aquatic ecology, 353–358
    on fish, 355–357, 373–374
  evaporation losses, 138, 145, 399
  modeling and prediction, 374–375, 389
  from nuclear and fossil power plants, compared, 344n, 352–353, 368–369, 397–398
    table, 398
  and power plant size, 345, 353, 394, 396–399
  and public power policy, 388–389
  reduction of, 217
  temperature rise from, 346, 349, 353–354
  uses for, 348, 358, 381–382, 388, 405–408
  water quality standards for, 347–348, 359–360, 372

Water
  for cooling, 344–346, 352, 360, 394, 406 (*see also* Cooling)
    tables, 345, 403
  recreational use, 351
Water chemistry, in reservoirs, 139–140
Water management
  in California, 163–164
  flow stabilization, 138, 146
  use of pumped storage for, 162–164
Water pollution
  by acid mine drainage, 320, 329–332
  methods and costs of control, 332–336
    table, 334
  by chemicals in cooling water, 404–405
Water quality
  in pumped-storage reservoirs, 168–169
  standards for thermal pollution, 347–348, 359–360, 372
Water-soil relationship
  near dams, 138, 142–143
  at pumped-storage reservoirs, 168–169
Water vapor, radiation absorption, 248, 252
Weapons tests, fallout, 35, 76
West European Atmospheric Chemistry Network, map of, 292
West Texas Utilities Company, 196
West Virginia, strip mining, 326
  reclamation techniques, photograph, 330
West Virginia Surface Mining Association, 330
Westinghouse Electric Company, 206, 217
Wheeler reservoir, 380
White, Gilbert F., 137
Widows Creek power plant, 110–111, 113, 116
  bone and lung dose rates from, 114
  fly ash and particulates from, 116
  radioactive releases and dose rates, table, 111

Index 440

Widows Creek power plant (continued)
  site characteristics, 116
Wilms' tumor, radiation-induced, 51
Wilson, Daniel W., 63–74
Winkler, F., 189–190
Wolff, Nigel O'C., 95
Wolk, Ronald H., 228–245
Woodrow crude oil, sulfur content of, table, 309
Wyoming crude oil, sulfur content of, table, 309

Xenon-133, 29
Xenon-135, 29, 37
Xenon-138, 29
X rays
  dose from diagnostic use of, 51–52, 96, 103
    lack of standards for, 96
    order of risk from, 101
  during pregnancy, 50, 69, 84–91
    table, 87
  repair of damage from, 67
  and reproductive performance, 82–91

Yankee Atomic power plant, 109
  noble and activation gases from, table, 115
  radioactive effluent from, tables, 36
Yards Creek Pumped Storage Generating Station, 163
  photograph, 160
Yukon River, 153

SAINT JOSEPH'S COLLEGE, INDIANA
TD172.5 .P68   ISJA
Power generation and environmental change; symposium of the

3 2302 00008 3222